EXERCISES IN ACTIVE TECTONICS

An Introduction to Earthquakes and Tectonic Geomorphology

NICHOLAS PINTER

Yale University

PRENTICE HALL, UPPER SADDLE RIVER, NJ 07458

Production editor: James Buckley
Production supervisor: Joan Eurell
Acquisitions editor: Robert McConnin
Production coordinator: Ben Smith

Simon & Schuster / A Viacom Company
Upper Saddle River, NJ 07458

Printed in the United States of America

10 9 8 7 6 5 4 3 2 1

ISBN 0-13-517905-X

Prentice-Hall International (UK) Limited, *London*
Prentice-Hall of Australia Pty. Limited, *Sydney*
Prentice-Hall Canada, Inc., *Toronto*
Prentice-Hall Hispanoamericana, S.A., *Mexico*
Prentice-Hall of India Private Limited, *New Delhi*
Prentice-Hall of Japan, Inc., *Tokyo*
Simon & Schuster Asia Pte. Ltd., *Singapore*
Editora Prentice-Hall do Brasil, Ltda., *Rio de Janeiro*

TABLE OF CONTENTS

PREFACE

Active tectonics is an area of intensive research, focusing on deformation of the Earth's crust and the hazards presented to society by earthquakes and faulting. It incorporates a range of disciplines: seismology, geomorphology, geodesy, structural geology, plate tectonics, and hazard planning. The National Science Foundation has named active tectonics as a new Special Emphasis Area, with the goal of promoting research that will mitigate the human and economic costs of tectonic events. The interdisciplinary nature of active tectonics makes it a challenging topic to cover in a university class, but students will be rewarded with material that is varied, interesting, timely, and vitally important to society.

This exercise manual provides a set of exercises for classes in the areas of tectonic geomorphology, earthquakes, and general geomorphology. The manual is intended to enrich lecture- and textbook-based curricula in any of these areas. The exercises focus on problem-solving, presenting students with real-world data and techniques, with step-by-step instructions and explanations. Because all the exercises are fully self-contained, they are suitable either as a comprehensive laboratory curriculum or as homework assignments. In addition, four *Regional Focus* chapters provide brief summaries, technical and non-technical references, and discussion questions suitable as starting points for classroom discussions or term papers on the tectonic framework of the U.S.

These exercises can be used to enrich a class that uses the new Active Tectonics text by Edward Keller and Nicholas Pinter, or in any class covering earthquakes or tectonics in detail. In addition, because these exercises focus on topographic maps, air photos, radiometric dating methods, fluvial systems, coastal systems, active folding, faulting, hillslope development, and regional landscape development, this manual can be used in any geomorphology class, particularly where the instructor wishes to provide a unifying theme for his or her class. Because the exercises are self-contained, this manual is appropriate for undergraduate students. At the same time, because the exercises outline cutting-edge techniques in active tectonics, advanced students and professionals should also find this manual useful.

These exercises originally were developed for a class in "Earthquakes and Landscape" given by the author at University of Vermont, where it supplemented the Keller and Pinter Active Tectonics text. The students in that class found several features of the exercises to be particularly rewarding:

- Application of textbook information
- Emphasis on problem solving: Each exercise is a unified problem that reinforces and applies one or more theoretical concepts in tectonic geomorphology.
- Focus on real research problems and techniques: Air photographs and maps show landforms created by faulting and active deformation. Problems show students how landforms are used to infer past tectonic activity and future seismic hazard. In most chapters, exercises use real-world examples and data from the recent scientific literature.
- All exercises are self-contained and fully workable: Each exercise includes an introduction, one or more examples, and all information necessary for completing the assignment. Each exercise has been used by a range of undergraduate geology majors and non-majors, and the manuscript has been modified to benefit from their input.

EXERCISE 1

LOCATING EARTHQUAKE EPICENTERS

Supplies Needed

- calculator
- metric ruler
- compass (circle-drawing variety)
- colored pencils (red, green, blue)

PURPOSE

Seismology is the study of the waves generated by earthquakes and transmitted through the Earth. To seismologists, seismic waves are like sonar on a submarine or radar on an airplane, allowing them to study the great volume of the Earth that lies hidden beneath the surface. Locating where earthquakes occur is one of the most basic applications of seismology, and that information is crucial in identifying and characterizing faults and regional fault zones. Using data from the 1994 Northridge earthquake, which caused about $20 billion in damage in the Los Angeles area, this exercise shows you how seismologists locate individual earthquakes. Using recent earthquake data from around the world, the exercise shows you how that data outlines the Earth's lithospheric plates and helps to characterize the motion occurring at the different plate boundaries.

INTRODUCTION

This book is about earthquakes, about how earthquakes shape the Earth's surface, and about using geomorphology (the geology of the surface) to infer past and future earthquake activity. An earthquake is defined as "a sudden motion or trembling in the Earth caused by the abrupt release of strain on a fault[1]." A fault is a break in the Earth's crust on which rupture occurs or has occurred in the past. Faults are classified according to the type of rupture that occurs on them (Figure 1.1). Motion on a *normal fault* is predominantly vertical and is caused by tension or extension. The block overlying the fault (the "hanging-wall block" in fault lingo) moves down relative to the block beneath the fault (the "footwall block"). Motion on a *reverse fault* also is predominantly vertical, but it is caused by compression, and the hanging-wall block is pushed up relative to the footwall

[1] after American Geological Institute, 1976. Dictionary of Geological Terms. Anchor Books.

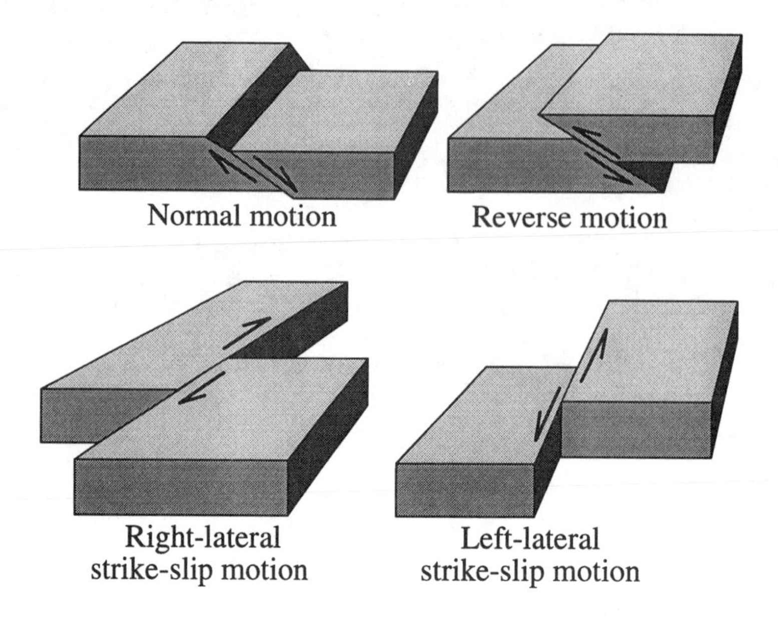

Figure 1.1. Types of faults based on sense of movement.

block. Strike-slip faults are characterized by horizontal motion, and material is displaced either to the right or to the left relative to material on the opposite side of the fault.

An earthquake is caused by the sudden release of elastic (recoverable) strain that gradually builds up on a fault over time. Strain accumulation may occur over just a few decades, such as on a major fault system like the San Andreas fault, or over thousands of years on slower-moving faults. In general terms, a fault ruptures when the amount of strain exceeds the strength of the rocks, but the actual triggering mechanism or process is not well known. The energy of the resulting earthquake depends on the amount of strain built up, the strength of the rocks along the fault, and the dimensions of the rupture area. Rupture begins at a single point on the fault surface, known as the earthquake's *focus,* but it spreads rapidly. The largest earthquakes may break a fault or faults over several hundred kilometers. When newspapers report the occurrence of an earthquake, the location they cite usually is the earthquake's *epicenter.* The epicenter is the point on the surface directly above the earthquake's focus (Figure 1.2).

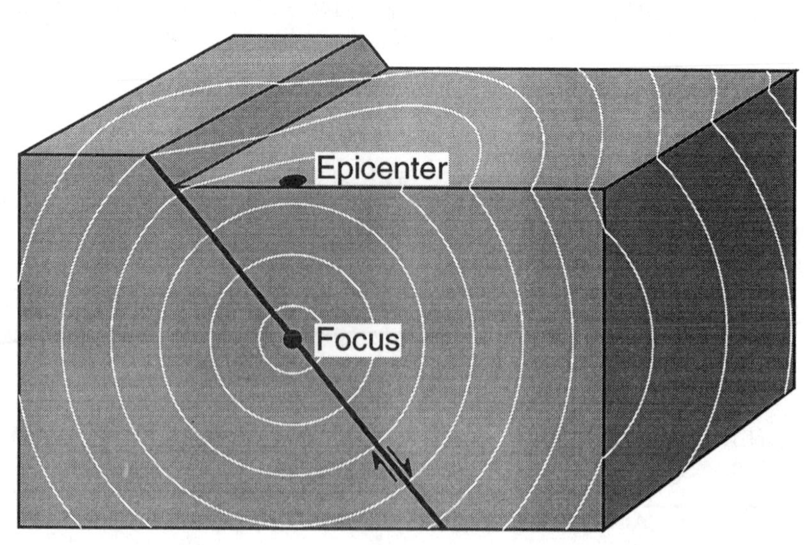

Figure 1.2. The focus and epicenter of an earthquake.

Perhaps the most important aspect of earthquakes, at least to people near the epicenter, is shaking. Shaking is the result of *seismic waves* that are transmitted to or along the Earth's surface. Rupture on a fault causes seismic waves, similar to the way a thrown stone causes ripples on the surface of a pond. There are three main types of seismic waves, and they are categorized by their type of motion (Figure 1.3). P-waves are compressional, so that particles displaced by the waves move forward and back parallel to the direction the wave propogates. S-waves are shear waves, in which particles move perpendicular to the the propogation direction. There are two types of surface waves (Love waves and Rayleigh waves), involving either shearing or elliptical motion.

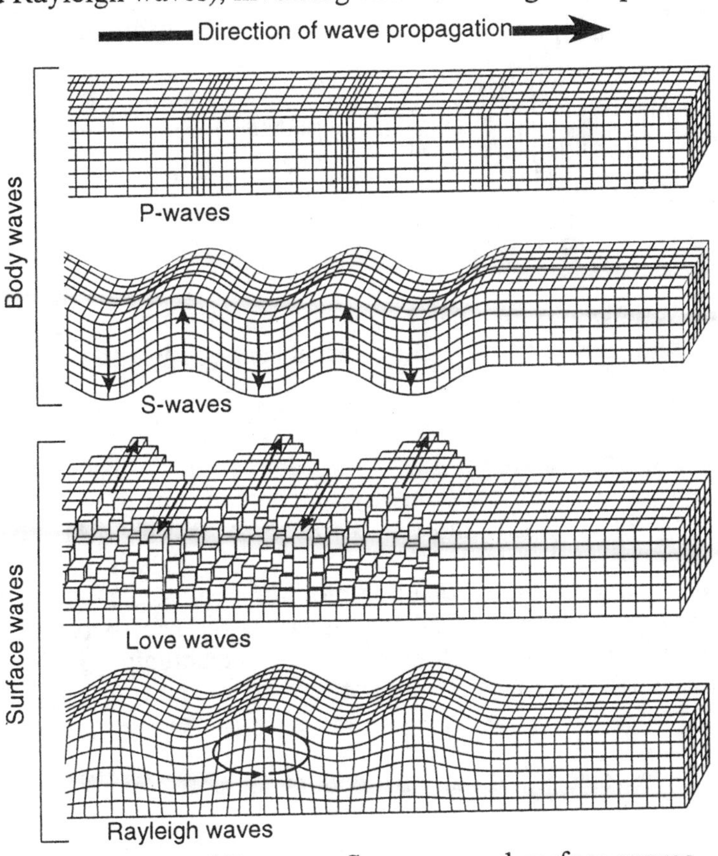

Figure 1.3. The nature of P-waves, S-waves, and surface waves. (After Bolt, 1988. Earthquakes. W.H. Freeman: New York)

LOCATING EARTHQUAKES

As mentioned earlier, seismic waves are the main tool of seismologists, allowing them to unravel the properties of faults and rocks deep beneath the surface of the Earth. Seismic waves propagate throughout the Earth, and can be detected both close to earthquake epicenters and on the opposite side of the planet. The instruments that seismologists use to detect seismic waves are *seismometers.* Early seismometers consisted simply of a rotating drum and a pen mounted on a free-swinging arm. When all was quiet, the pen would draw a straight line on the drum as it rotated, but when seismic shaking occurred, the pen would create a graphical image of the passing seismic waves. This graphical image is called a *seismogram,* and one is illustrated in Figure 1.4.

Locating Earthquake Epicenters

An important detail about the different types of seismic waves is that each type travels at a different speed. P-waves travel the fastest, S-waves not as fast, and surface waves more slowly. In fact, the "P" in "P-wave" stands for "primary" because they are the first waves to arrive after an earthquake. The "S" in "S-wave" stands for "secondary" because they arrive after the P-waves. The different travel times of seismic waves are the key to locating the epicenters of earthquakes.

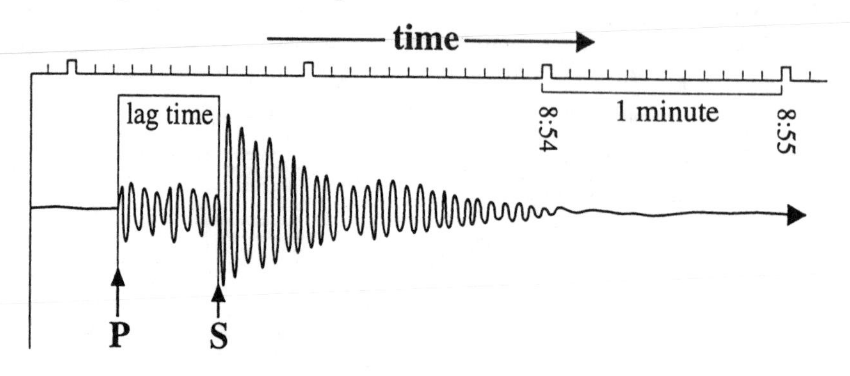

Figure 1.4. A seismogram. The first arrivals of the P-waves and S-waves are shown. The lag time is the interval between P- and S- wave arrivals.

Example 1.1.

To understand how seismograms are used to find the distance to an earthquake epicenter, imagine that car A and car B always depart for trips together, but car A always travels at 100 km per hour (kph), and car B travels at 85 kph. An observer anywhere along the cars' route could calculate exactly how far they had traveled simply by measuring the time between them. For example, if the car A passes a given spot at 2:30 pm, and car B passes the same spot at 2:45 pm, then the distance between that spot and the cars' point of departure must be about 142 km. This calculation is simply the result of knowing that distance traveled (d) is the product of rate (r) and time (t):

$$d = r * t \tag{1.1}$$

Because the distance traveled is the same for both cars, the following must be true:

$$d = r_A * t_A = r_B * t_B \tag{1.2}$$

Given the speed of the two cars (r_A and r_B) and that car B passed the spot 15 minutes after car A ($t_B = t_A + 0.25$ hrs), Equation 1.2 becomes:

$$100 \text{ km/hr} * t_A = 85 \text{ km/hr} * (t_A + 0.25 \text{ hrs}) \tag{1.3}$$

Simplifying and solving for t_A:

$$t_A = 1.42 \text{ hr} \tag{1.4}$$

Combining Equations 1.4 and 1.2:

$$d = r_A * t_A \tag{1.5}$$

$$d = 100 \text{ km/hr} * 1.42 \text{ hr} \tag{1.6}$$

$$d = 142 \text{ km} \tag{1.7}$$

P-waves and S-waves are much like the two cars in Example 1.1. They both depart together (from the focus), and one travels consistently faster than the other. P- and S-wave velocities vary somewhat depending on the local geology, but they are consistent enough that seismic-wave *travel-time curves* (Figure 1.5) can be used for earthquakes and seismograms around the world. Travel-time curves are graphical solutions to the "distance equals rate times time" equation. Figure 1.5A is a curve for epicenters hundreds or thousands of kilometers away from the recording stations, while Figure 1.5B is an enlargement for distances of tens to a few hundreds of kilometers. Note that the P- and S-wave arrivals on Figure 1.5A are curved lines because the seismic waves travel through the interior of the Earth, while distance is measured along the surface.

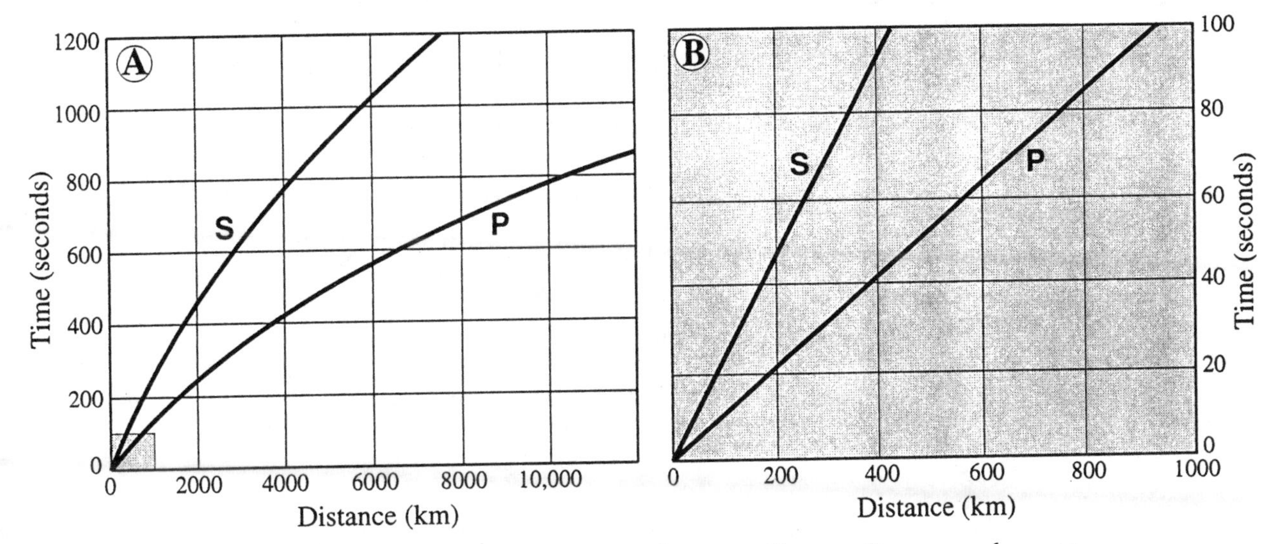

Figure 1.5. Seismic-wave travel times for recording stations anywhere on Earth (A) and less than 1000 km from the epicenter (B). Note that graph B is just an enlargement of the small shaded box in graph A.

To use a seismogram to calculate the distance between the recording station and the epicenter of an earthquake, follow these steps:

A) Identify the first arrival of the P-waves as shown in Figure 1.4.

B) Identify the first arrival of the S-waves.

C) Measure the lag time between the P-wave arrival and the S-wave arrival in seconds ($t_S - t_P$).

D) Find the distance that corresponds to that lag time using a travel-time curve (Figure 1.5).

Using the information above, answer the questions on the following page:

Locating Earthquake Epicenters

1) Using Figure 1.5B, calculate the average velocity of P-waves and of S-waves in the Earth's crust.

2) Go back to Figure 1.4. Determine the distance from the station that recorded that seismogram to the epicenter that caused those seismic waves. Note that the tic-marks on the time scale are in 5-second increments.

3) If you know that the first one-minute mark on Figure 1.4 is 8:52:00 p.m., find the exact time at which this earthquake occurred. [Hint: This problem becomes simple if you measure time as the number of seconds after or before some arbitrary time, for example 8:52:00. For example, 8:53:12 would be 72 sec, while 8:51:12 would be -48 sec. Only the final answer (the time of the earthquake) needs to be converted back into clock time.]

Being able to find distance from a seismogram is the first step in locating earthquake epicenters. For example, if you know that a seismometer in Berkeley, California is 6000 km from an earthquake, you would be able to draw a circle with a radius of 6000 km around Berkeley (Figure 1.6). In order to pinpoint that epicenter, however, you would need information from at least two additional recording stations, for example Honolulu, Hawaii and San Juan, Puerto Rico on Figure 1.6. The more stations you have data from, the more accurate is your estimate of the epicenter location.

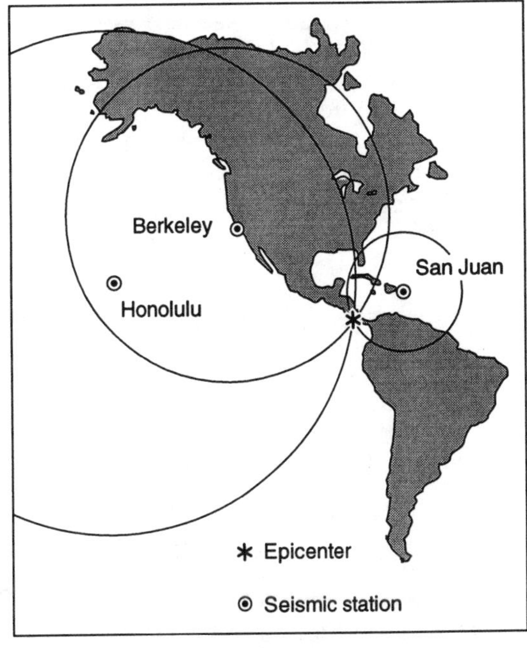

Figure 1.6. Locating an epicenter. (After Bolt, 1978)

Not every seismograph station around the world detects every earthquake. Some stations are located in *shadow zones* relative to a given epicenter, meaning that either P-waves, S-waves, or both do not reach those locations. Shear waves, for example, cannot pass through a fluid. The S-wave shadow zone is the main piece of evidence that led seismologists to conclude that the Earth has a fluid outer core. The exercise that follows uses data from seismographs relatively close to the earthquake epicenter in question, so that shadow zones and other complexities of a three-dimensional Earth do not present problems.

THE 1994 NORTHRIDGE EARTHQUAKE

In the pre-dawn hours of January 17, 1994, a magnitude M_w=6.7 earthquake struck Northridge, California, an urbanized area about 30 km northwest of downtown Los Angeles. As one eyewitness reported:

> *"Very loud noise during the shaking from floor boards and house moving and bricks falling from the chimney (narrowly missing my car) and contents of shelves, closets, and cupboards coming down on floor. I felt endangered enough after about five seconds of shaking to dive ... into a doorjamb for the duration of the quake. Stove, refrigerator, shelves moved one to two and a half feet from their original positions. A four drawerfile cabinet fell over. All fallen or moved items were displaced along a more or less East-West direction. Electricity, and phone immediately went off and stayed off for hours. A waterpipe leading to our home sprang a leak and water spurted out all over our back area."* (Quoted in Dewey et al, 1995)

The earthquake caused damage over much of the Los Angeles metropolitan region, including the destruction of several apartment complexes in the Northridge area and the collapse of three freeway overpasses that snarled L.A. traffic for months. In all, the Northridge earthquake killed 33 people, injured 7,000, severely damaged at least 24,000 structures, and caused an estimated $20 billion in damage, making it the most costly natural disaster ever (although the damage figure was surpassed just one year later by the Kobe earthquake in Japan).

Locating Earthquake Epicenters

The Northridge earthquake occurred near the "Big Bend" of the San Andreas Fault (see *Regional Focus A* later in this book) in an area characterized by complex compressional faulting and folding. The earthquake occurred on a thrust fault that was previously unknown because the fault does not break the surface. Fault rupture at depths of 8 km and below uplifted a broad area of the ground surface by up to 70 cm. The Northridge earthquake was a devastating wakeup call to many scientists who previously believed that the main seismic hazard in the Los Angeles region came from the San Andreas and other right-lateral strike-slip faults in the area.

THE SOUTHERN CALIFORNIA SEISMOGRAPH NETWORK

In this part of the exercise, you will use data from seismographs across Southern California to identify the location and precise time of the 1994 Northridge earthquake. In Figure 1.7, you will find seven seismograms from seven different stations recorded on January 17, 1994. Station codes are as follows:

BAR Barrett Dam
GSC Goldstone
NEE Needles
PAS Pasadena
PFO Pinion Flat
SVD Seven Oaks Dam
VTV Victorville

The zero-seconds mark on the time scale is 4:31:00 Pacific Standard Time for all the seismograms. For each seismogram, identify the first P-wave arrival and the first S-wave arrival. Enter this information in Table 1.1, including P-wave arrival time (t_p), S-wave arrival time (t_s), and P-S lag time ($t_s - t_p$). Use Figure 1.8 (an enlarged copy of Figure 1.5B) to find the distance between each station and the epicenter. The seismogram from BAR is already completed for you as an example. Enter those distances in Table 1.1.

Table 1.1. Summary of data from the Southern California seismograph network.
(t_P is the arrival time (in sec) of the first P-wave; t_S is the S-wave arrival time)

	t_P (sec)	t_S (sec)	$t_S - t_P$	distance (km)
BAR	35	64	29	245
GSC				
NEE				
PAS				
PFO				
SVD				
VTV				

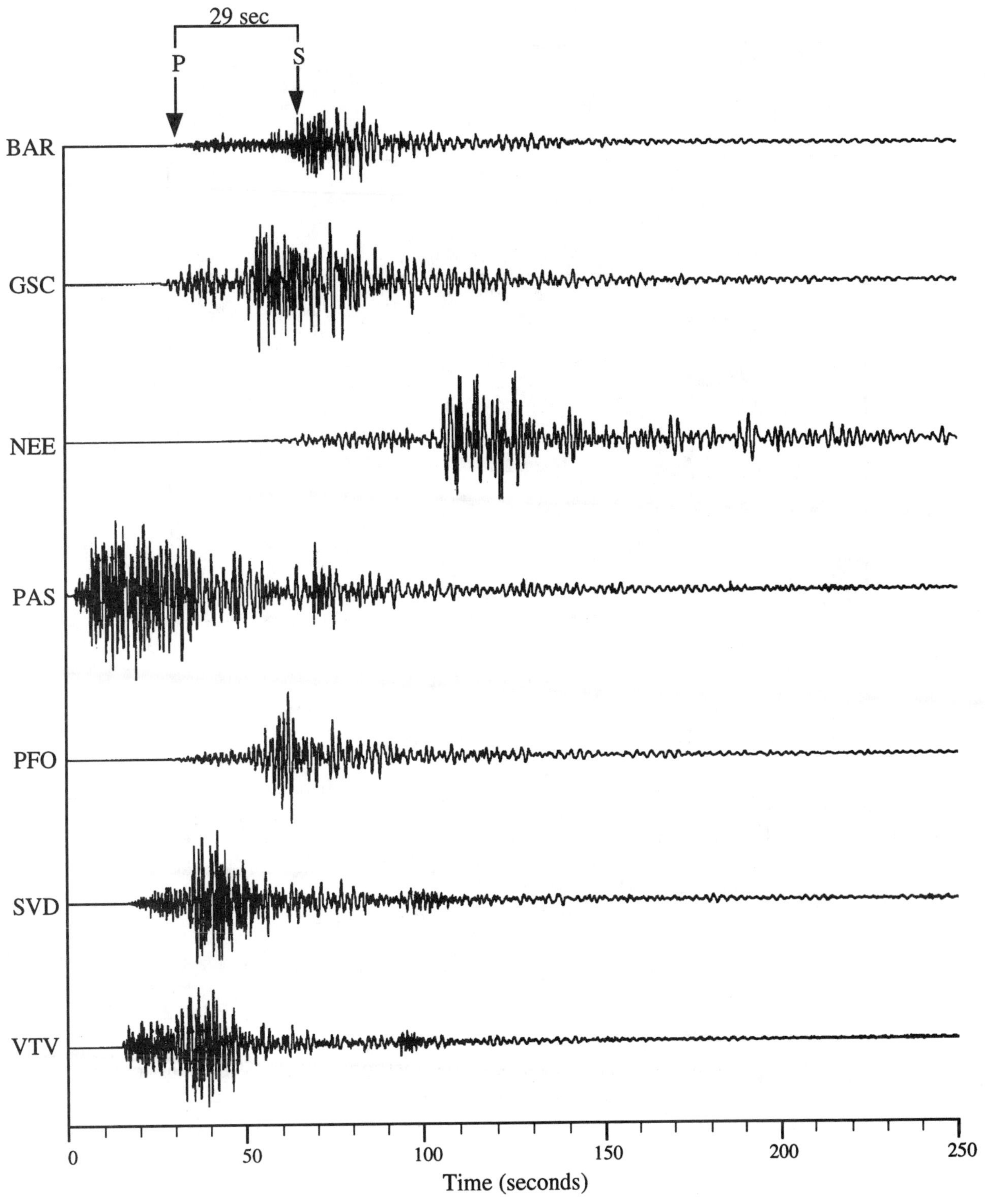

Figure 1.7. Seismogram records of the 1994 Northridge earthquake. P- and S-wave arrivals and the S–P lag time have already been completed for BAR as an example.

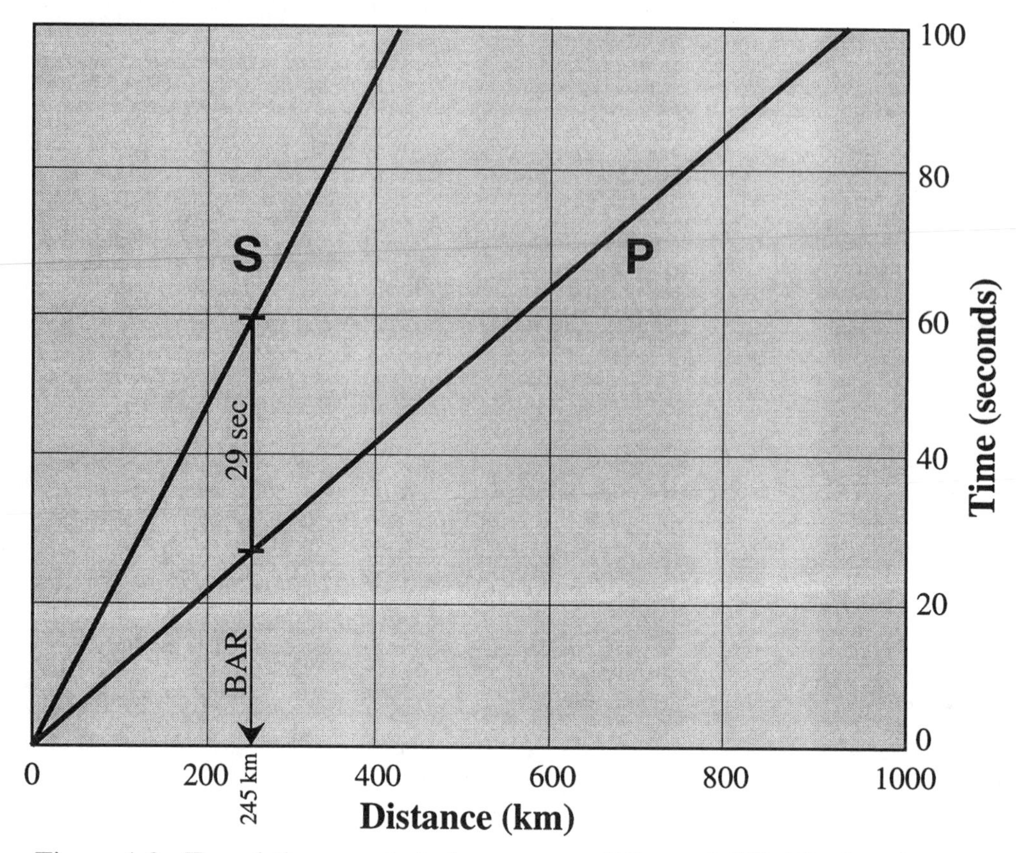

Figure 1.8. Travel-time graph (enlargement of Figure 1.5B). The lag time from station BAR is shown as an example.

After completing Table 1.1, use a compass to draw a circle around each seismograph station on Figure 1.9 corresponding to the distance from each station to the epicenter. Note that the map scale is in the lower left corner of the figure. The intersection of the seven circles is the location of the epicenter of the Northridge earthquake.

4) What are the latitude and longitude of the Northridge epicenter?

5) Pick one or two of the seismograms to determine exactly what time the earthquake began (hour, minute, and seconds). Use the method you used for Questions 1.1-1.3. The more seismograms you use, the more accurate your result.

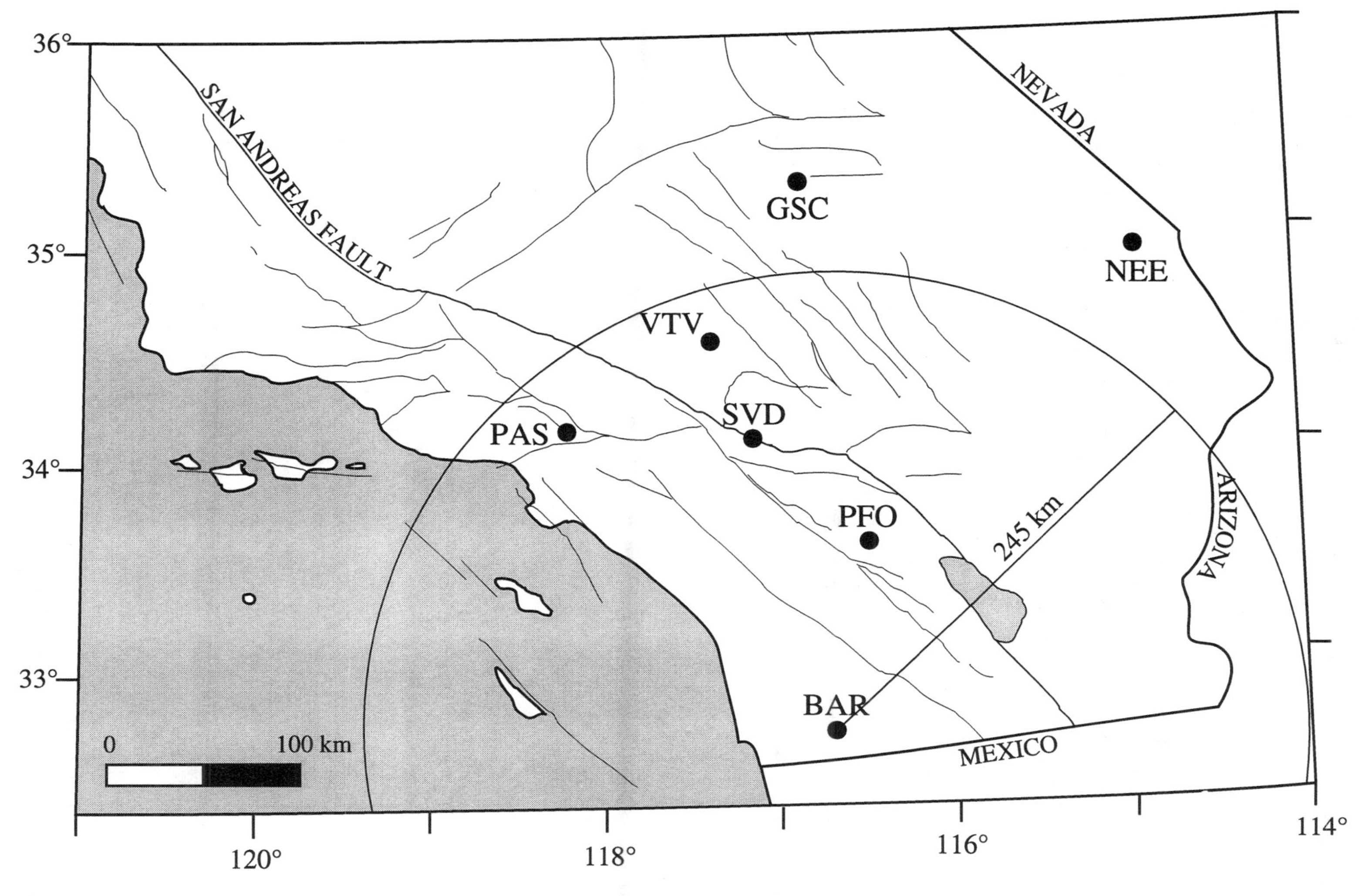

Figure 1.9. Location map of Southern California. Seismograph stations and major faults are also shown.

GLOBAL TECTONICS

The science of geology underwent a revolution in the 1960s. Geologists discovered that a vast number of geological phenomena that were previously believed to be unrelated were actually part of a unified global system of *plate tectonics.* The theory of plate tectonics was developed by identifying patterns in fields as different as volcanology, structural geology, marine geology, seismology, and paleontology. In a nutshell, plate tectonics is the theory that the Earth's surface is subdivided into distinct *plates* that move relative to the plates around them. Most earthquake activity, volcanism, and deformation of the crust is concentrated at the edges of the plates, where they interact. A global map of earthquake epicenters worldwide (Figure 1.10), for example, clearly shows the outlines of the major plates of the Earth.

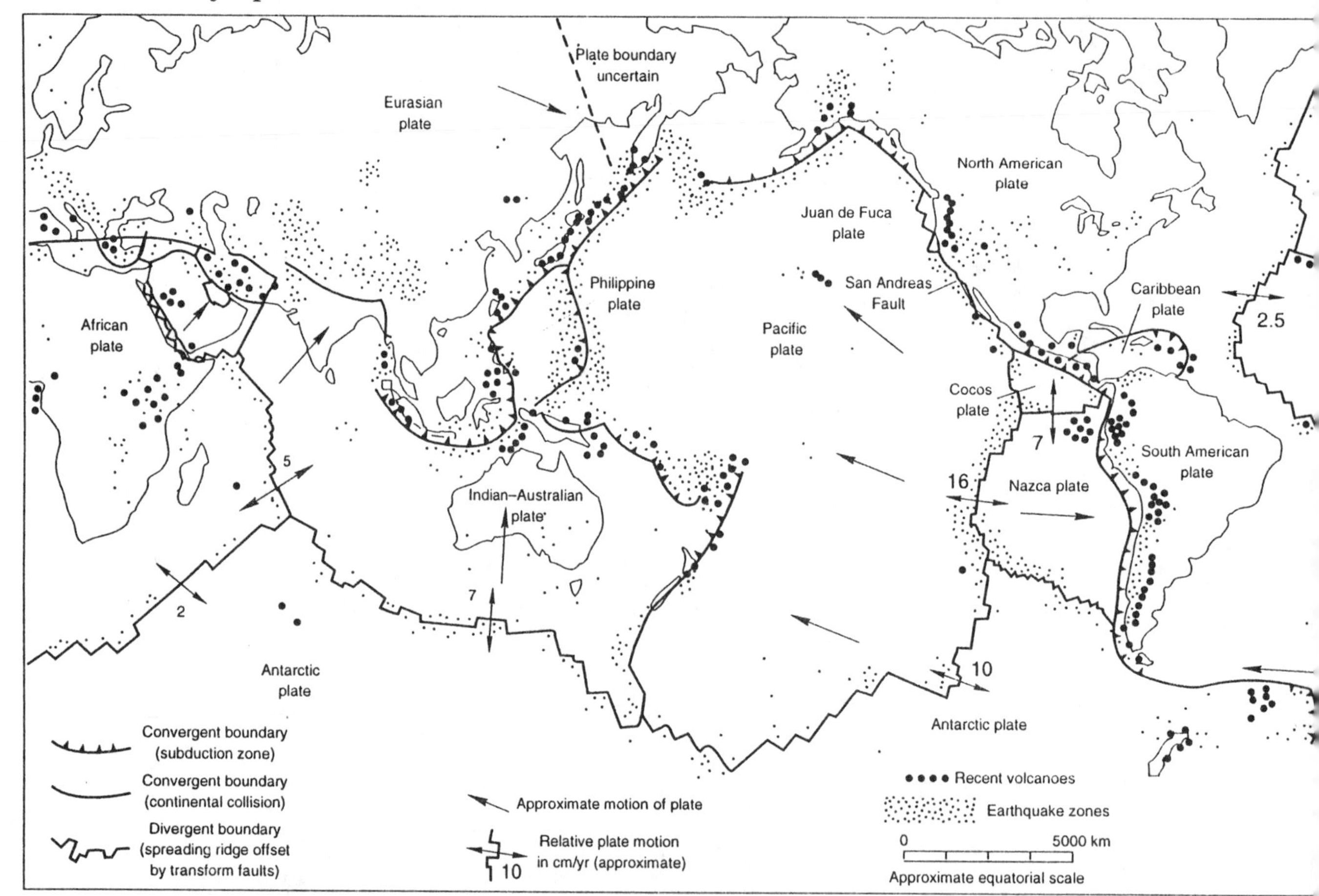

Figure 1.10. Map of global earthquake activity, 1963-1988 (Richter magnitude ≥ 5). (Courtesy of National Earthquake Information Center)

There are three basic types of plate boundaries: convergent boundaries, where the two plates move towards each other; divergent boundaries, where the plates move apart; and transform boundaries, where the plates move horizontally past one another. These plate-boundary types and their major geological characteristics are shown in Figure 1.11.

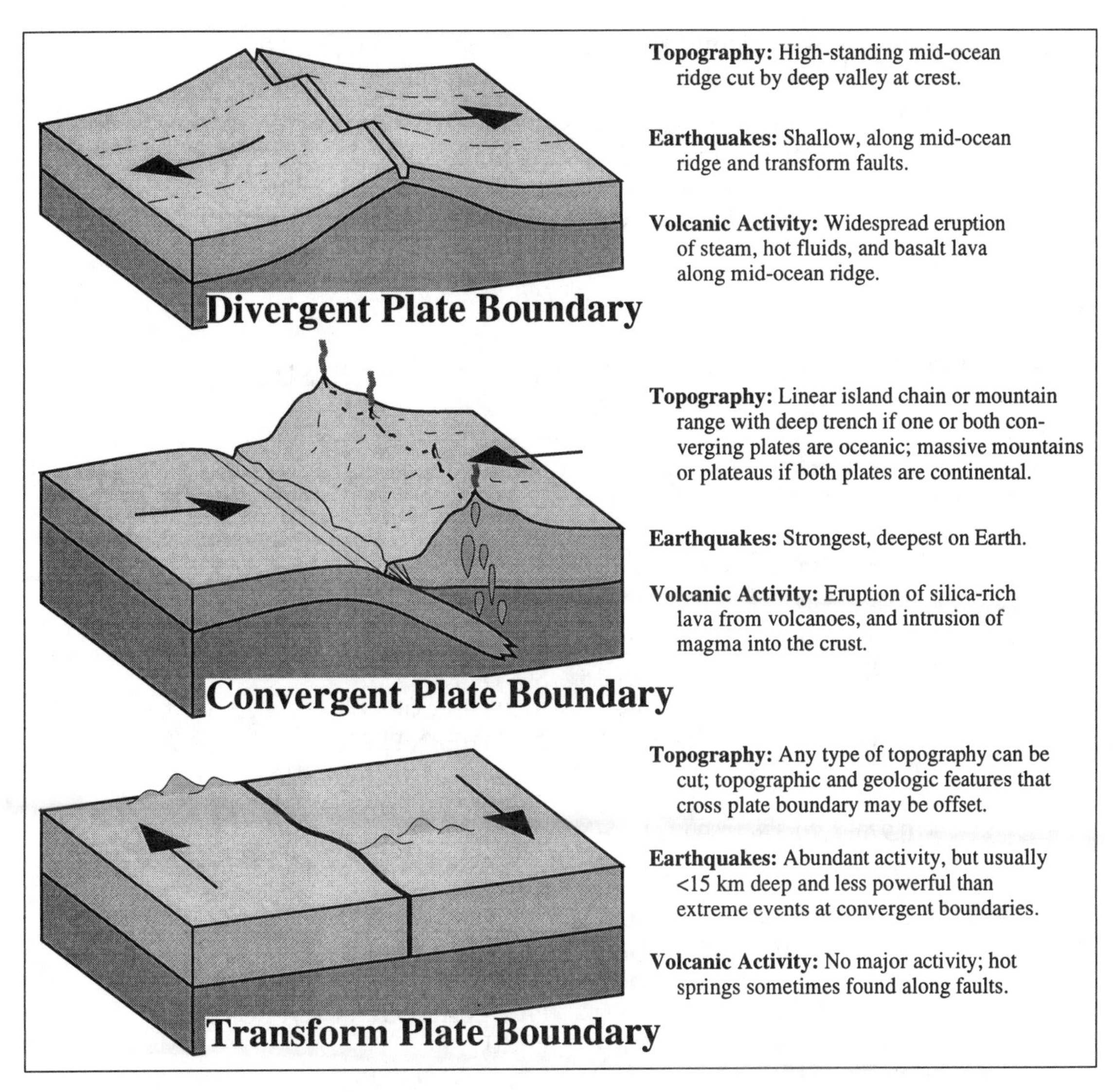

Figure 1.11. The three basic types of plate boundaries and their major geologic characteristics.

The data set on the following page is a list of earthquakes compiled by the U.S. Geological Survey from 1990-1995, including the latitudes and longitudes of the epicenters, focus depths, magnitudes, and a general statement of the locations. Each year, many thousands of earthquakes occur around the world. The earthquake list here is limited to events with magnitudes of 6.5 or more and earthquakes that caused fatalities or significant damage. The list is further limited to the geographical area shown in Figure 1.12 (Central and South America).

The purpose of this next exercise is to show how epicenter information can be used to infer regional patterns of plate-tectonic activity. The information that you have available here is similar to the information that geologists used to help develop the theory of plate tectonics in the 1950s and 1960s. Follow these steps:

Locating Earthquake Epicenters

A) Using the latitude and longitude information, find the location of each earthquake's epicenter on Figure 1.12.

B) Using the depth information, plot each epicenter location using the following guide:

0-50 km depth: red dot
51-200 km depth: green dot
>200 km depth: blue dot

C) Answer the questions that follow.

SIGNIFICANT EARTHQUAKES OF CENTRAL AND SOUTH AMERICA, JANUARY, 1990 – APRIL, 1995

+ latitude is degrees north of the equator, – latitude is degrees south; longitude is degrees west of Greenwich; depth is in km.

	LAT	LON	DEP	MAG	REGION
1990	9.919	84.808	22	M_s 6.8	COSTA RICA
	11.426	86.301	15	M_s 6.7	NICARAGUA
	6.905	82.622	10	M_s 6.5	SO. OF PANAMA
	-6.016	77.229	24	M_s 6.6	NORTHERN PERU
	-6.062	77.136	26		NORTHERN PERU
	12.925	87.723	22		COAST OF NICARAGUA
	-0.059	78.449	5		ECUADOR
	-10.970	70.776	599	M_b 6.8	PERU-BRAZIL BORDER
	9.869	84.302	17	M_s 6.1	COSTA RICA
1991	-6.038	77.130	21	M_s 6.4	NORTHERN PERU
	-5.982	77.094	20	M_s 6.7	NORTHERN PERU
	9.685	83.073	10	M_s 7.4	COSTA RICA
	9.542	82.418	10	M_s 6.2	PANAMA-COSTA RICA BORDER
	-13.108	72.187	105	M_b 6.5	PERU
	-15.679	71.574	5		SOUTHERN PERU
	4.554	77.442	21	M_D 6.8	COAST OF COLOMBIA
1992	10.210	84.323	79	M_D 5.6	COSTA RICA
	11.742	87.340	45	M_s 7.4	NICARAGUA
	6.866	76.816	10	M_s 7.0	NORTHERN COLOMBIA
	7.123	76.887	10	M_s 7.4	NORTHERN COLOMBIA
1993	-11.652	76.530	106	M_w 6.3	CENTRAL PERU
	-31.560	69.234	113	M_w 6.3	ARGENTINA
	9.821	83.622	20	M_w 5.8	COSTA RICA
	-25.304	70.166	48	M_w 6.6	NORTHERN CHILE
1994	-13.339	69.446	596	M_w 6.9	PERU-BOLIVIA BORDER
	-28.299	63.252	562	M_w 6.9	ARGENTINA
	-28.501	63.096	601	M_w 6.9	ARGENTINA
	7.414	72.033	12	M_w 5.9	NORTHERN COLOMBIA

	2.917	76.057	12	Mw 6.7	COLOMBIA
	-13.841	67.553	631	Mw 8.2	NORTHERN BOLIVIA
	-26.642	63.421	564	Mw 6.5	ARGENTINA
1995	5.075	72.918	18	Mw 6.5	COLOMBIA
	4.162	76.644	69	Mw 6.4	COLOMBIA
	1.289	77.303	5	Mw 4.4	COLOMBIA
	-3.854	76.958	103	Mw 6.7	NORTHERN PERU

6) Looking at all of your epicenters plotted on Figure 1.12, where do the deepest earthquakes occur relative to the more shallow earthquakes? Explain why this is the case (refer to Figure 1.11 if you need to).

7) Imagine that you are the first geologist to have accurate epicenter-location information. Take a pen or pencil and draw where the major plate boundary or boundaries seem to be located on Figure 1.12.

8) Last of all, consider that the list of earthquakes you used is *not* a complete list of all the earthquakes that occurred. Review the list of criteria used to limit that list. Are there any systematic biases in that list? (For example, are there some locations where you may have plotted a greater portion of all the earthquakes that occurred than in other areas?)

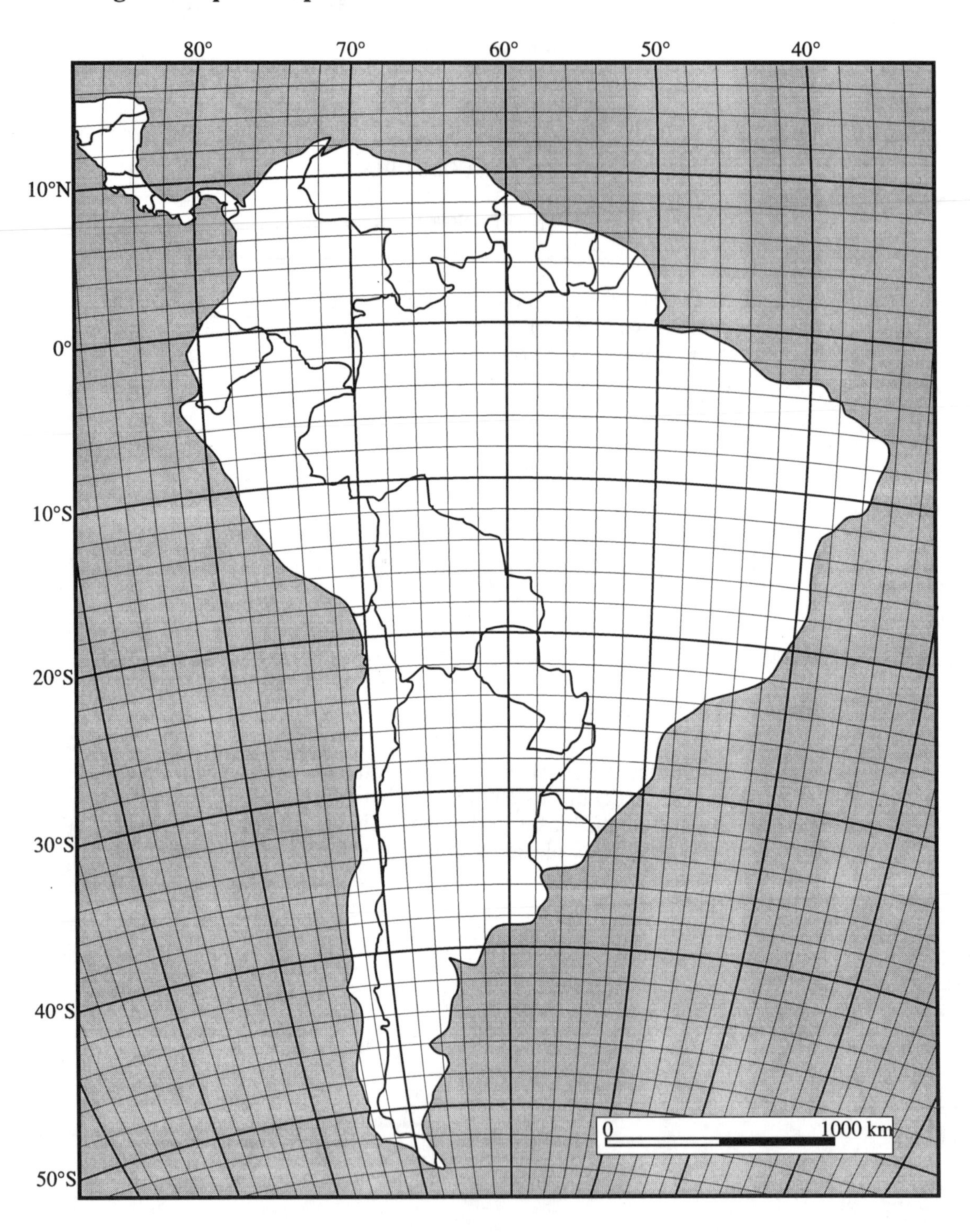

Figure 1.12. Location map of Central and South America.

BIBLIOGRAPHY

Benz, H.M., and J.E. Vidale, 1993. Probing Earth's interior using seismic arrays. Geotimes, 38(7): 20-22.

Bolt, B.A., 1978. Earthquakes. W.H. Freeman: San Francisco.

Bolt, B.A., 1993. Earthquakes (3rd Edition). W.H. Freeman: San Francisco.

Bott, M.H.P., 1982. The Interior of the Earth: Its Structure, Constitution, and Evolution. Edward Arnold: London.

Dewey, J.W., B.G. Reagor, L. Dengler, and K. Moley, 1995. Intensity distribution and isoseismal maps for the Northridge, California earthquake of January 17, 1994. U.S. Geological Survey Open-File Report 95-92.

Heppenheimer, T.A., 1987. Journey to the center of the earth. Discover, 8 (11): 86-90+.

Jones, L., and 30 others, 1994. The magnitude 6.7 Northridge, California earthquake of 17 January, 1994. Science, 266: 389-397.

Scholz, C.H., 1990. The mechanics of earthquakes and faulting. Cambridge University Press: Washington, DC.

Acknowledgements: The author would like to gratefully acknowledge the assistance of Paul Roberts of the California Institute of Technology, who assembled and prepared the seismograms and other data for this exercise. Earthquake epicenter data was obtained from the U.S. Geologi-cal Survey Earthquake and Geomagnetic Information on-line information service.

EXERCISE 2

EARTHQUAKE MAGNITUDE AND INTENSITY

Supplies Needed

- calculator
- metric ruler

PURPOSE

Exercise 1 introduced many of the fundamental concepts of earthquakes, and this exercise will add two more important concepts: earthquake magnitude and intensity. Both magnitude and intensity are expressions of the amount of energy released when a fault ruptures. Scientists are interested in measuring seismic energy in order to categorize earthquakes and to better understand tectonic processes. Society is interested in the strength of past and future earthquakes in order to assess and predict damage and loss of life. The following exercise will use data from the 1994 Northridge earthquake to illustrate the method for calculating Richter magnitude and for mapping seismic-shaking intensity.

MAGNITUDE

Magnitude is a measurement of the energy released by an earthquake. The first earthquake-magnitude scale was the *Richter scale,* devised by Charles F. Richter, a seismologist at the California Institute of Technology. The Richter scale is based on the amplitude of seismic waves – the stronger the earthquake, the stronger the seismic vibrations it causes. The Richter magnitude of an earthquake is expressed as a decimal number, such as 6.7. The most important thing to remember about Richter magnitude is that it is a logarithmic scale, meaning that an increase of *one* in magnitude corresponds to a factor of *ten* increase in the amplitude of ground motion. For example, a magnitude 6.7 earthquake causes shaking 10 times greater in amplitude than a magnitude 5.7 earthquake and 100 times greater than a magnitude 4.7 earthquake.

Mathematically, an earthquake of magnitude x results in seismic waves with amplitudes proportional to 10^x. The actual seismic-wave amplitude at a particular site depends on the distance of the site from the earthquake epicenter, the depth of the earthquake, and local near-surface conditions. Example 2.1 shows you how to compare the shaking that results from earthquakes with different magnitudes.

Magnitude and Intensity

Example 2.1.

Compare the seismic shaking produced by a magnitude 8.2 earthquake with the shaking from a magnitude 6.7 earthquake?

When we say that the amplitude of seismic waves (A) from an earthquake of magnitude x are proportional to 10^x, that is equivalent to saying:

$$A_{(M=x)} = k * 10^x$$

where k is an arbitrary constant. Thus:

$$A_{(M=8.2)} = k * 10^{8.2}$$

and:

$$A_{(M=6.7)} = k * 10^{6.7}$$

When asked to compare the shaking produced by two earthquakes, you are being asked to solve for the ratio of the two earthquake amplitudes, in this case $A_{(M=8.2)} \div A_{(M=6.7)}$:

$$A_{(M=8.2)} \div A_{(M=6.7)} = k * 10^{8.2} \div k * 10^{6.7}$$

$$k * 10^{8.2} \div k * 10^{6.7} = 10^{8.2} \div 10^{6.7} = 10^{(8.2-6.7)} = 10^{1.5}$$

thus: $$A_{(M=8.2)} \div A_{(M=6.7)} = 10^{1.5} = 31.6$$

Answer: A magnitude 8.2 earthquake creates shaking 31.6 times more greater in amplitude than a magnitude 6.7 earthquake.

1) The 1906 San Francisco earthquake had a magnitude of about 8.3. The 1989 Loma Prieta earthquake that struck San Francisco had a magnitude of 7.1. How much greater was the shaking in 1906 earthquake compared with the shaking in 1989?

2) How much greater was the shaking during the 1906 San Francisco earthquake than the shaking during a magnitude 4.0 tremor?

The method for determining the magnitude of an earthquake is illustrated in Figure 2.1 below. Richter magnitude (M) is a function of the amplitude of the largest wave on a seismogram and the distance from the recording station to the epicenter (measured either directly in kilometers or indirectly as the S–P lag time; see Exercise 1). On Figure 2.1, the magnitude is determined by connecting the maximum wave amplitude (85 mm with proper scaling of the seismogram) with the epicentral distance (300 km, or 34 sec S–P lag). The magnitude of the earthquake shown is the intersection of that line with the magnitude axis of the diagram at M=6.0.

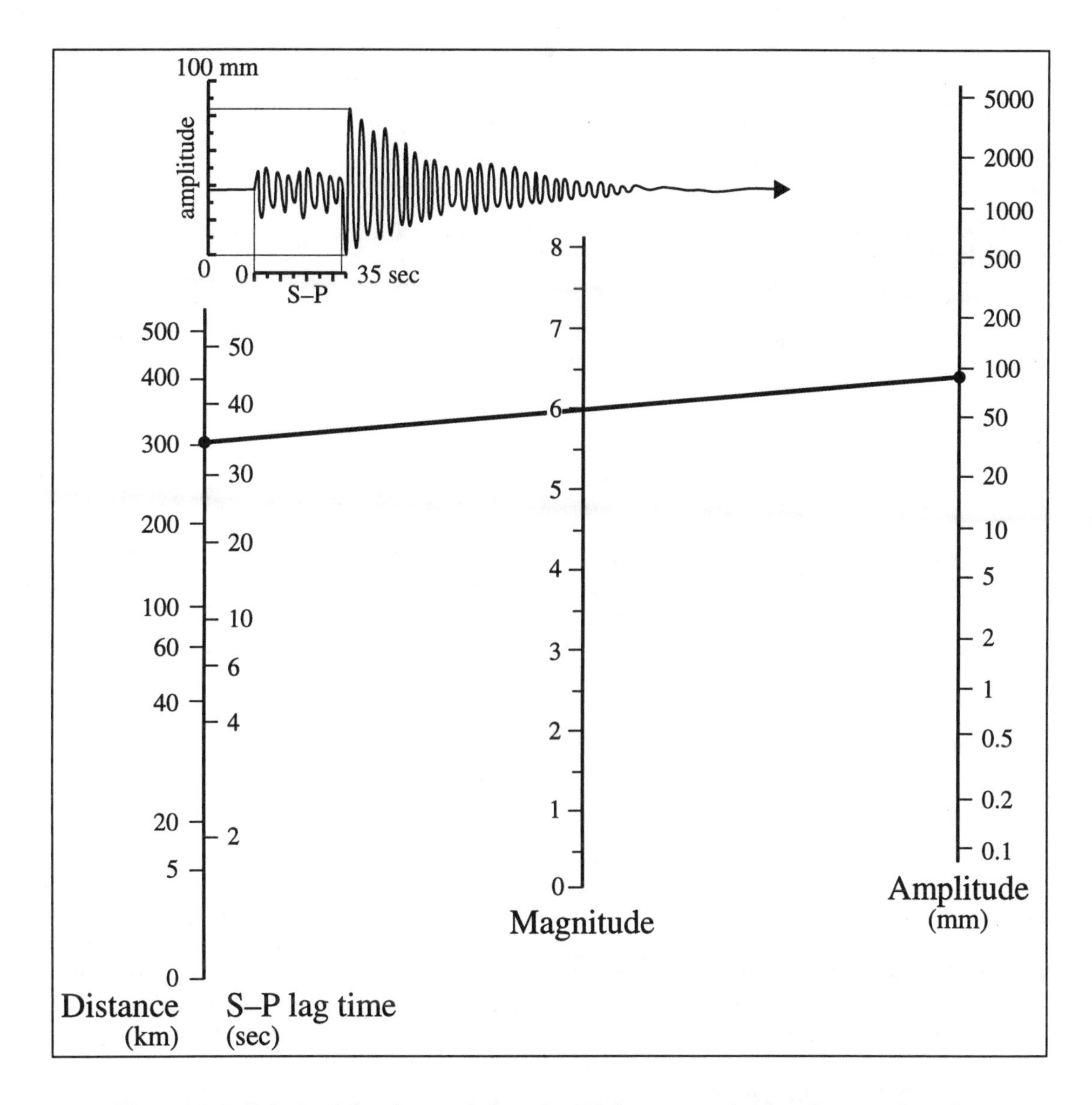

Figure 2.1. Method for determining the Richter magnitude of an earthquake from a seismogram. The maximum wave amplitude on the seismogram is connected with the epicentral distance. The intersection of that line with the magnitude axis gives the earthquake magnitude. (After Bolt, 1978)

Magnitude and Intensity

3) Using Figure 2.1, determine the Richter magnitudes for the earthquake data given in the table below.

Amplitude (mm)	Distance (km)	Richter Mag.
1	100	_____
10	100	_____
100	100	_____
10	5	_____
10	50	_____
10	500	_____

4) Looking at the first three earthquakes in the table above, what is the effect on Richter magnitude of a ten-fold increase in maximum seismic-wave amplitude? Why does this occur?

5) Looking at the last three earthquakes in the table above, what is the effect on Richter magnitude of a ten-fold increase in the distance between the recording station and the epicenter? Why does this occur?

It's important to note that Richter magnitude is not the only system for measuring earthquake energy. You may have noted that the list of earthquakes at the end of Exercise 1 included magnitudes with a variety of subscripts: M_s, M_b, M_w, and M_D. All of these magnitude scales are logarithmic scales, but different systems have advantages in different situations. For example, whereas Richter magnitude is based on the amplitude of the largest wave associated with a given earthquake, magnitude also can be determined using the largest body wave (M_b) or the largest surface wave (M_s). If surface waves cause a particular seismograph to go "off scale" (the amplitude is greater than the seismograph's range of motion), then the smaller body waves may still be used to calculate magnitude.

One particularly useful alternative to Richter magnitude is the *Moment magnitude scale* (M_w). Moment magnitude is based on the *seismic moment* of an earthquake, which is a direct measurement or estimate of the energy released by the earthquake. Seismic moment (M_o) can be calculated as follows:

$$M_o = \mu * D_{av} * A \qquad (2.1)$$

where μ is the modulus of rigidity of the crust (about $3.3 * 10^{11}$ dynes/cm^2; Brune, 1968), D_{av} is the average displacement on the fault during the earthquake, and A is the total area of rupture on the fault. Seismologists often favor using seismic moment because it is the most physically-based estimate of earthquake energy. Seismic moment can be converted into a magnitude scale using the following equation (Hanks and Kanomori, 1979):

$$M_w = 2/3 * \log M_o - 10.7. \qquad (2.2)$$

Example 2.2.

Find the seismic moment (M_o) of a M_w=7.5 earthquake.

The answer to this question is a straightforward solution of Equation 2.2:

$$7.5 = 2/3 * \log M_o - 10.7$$

simplifying:

$$3/2 * (7.5 + 10.7) = \log M_o$$

$$\log M_o = 27.3$$

The inverse of the logarithm function is the exponent function (10^x). The way to simplify a logarithm term is with the rule that $10^{(\log x)} = x$, so that:

$$10^{(\log Mo)} = 10^{(27.3)}$$

$$M_o = 10^{27.3}$$

$$M_o = 2.00 * 10^{27}$$

and the units of seismic moment are dyne·cm.

If the M_w=7.5 earthquake discussed above ruptures the surface with an average displacement (D_{av}) of 2.5 m, find the fault area (A) that ruptured during the earthquake.

This question uses the result from the last question ($M_o = 2.00 * 10^{27}$) and Equation 2.1:

$$M_o = \mu * D_{av} * A$$

$$2.00 * 10^{27} \text{ dyne·cm} = 3.3\ 10^{11} \text{ dynes/cm}^2 * 2.5 \text{ m} * A$$

Converting the meters term into cm and simplifying:

$$2.00 * 10^{27} \text{ dyne·cm} = 3.3 * 10^{11} \text{ dynes/cm}^2 * 250 \text{ cm} * A$$

$$2.4 * 10^{13} \text{ cm}^2 = A$$

$$A = 2.4 * 10^9 \text{ m}^2$$

$$A = 2.4 * 10^3 \text{ km}^2 = 2400 \text{ km}^2$$

MAGNITUDE OF THE 1994 NORTHRIDGE EARTHQUAKE

Exercise 1 introduced the Northridge earthquake just northwest of Los Angeles. This exercise will expand upon that example. It's important to note that, although it was the most damaging earthquake in U.S. history, the Northridge earthquake was definitely not the "Big One" that California fears. In fact it was over six times less powerful than the 1992 Landers earthquake that did relatively little damage. The major problem was that the Northridge earthquake struck in the middle of a densely-populated urban area.

6) Determine the Richter magnitude of the Northridge earthquake. Use the following steps:

A) Measure the S-P lag time for each seismogram.

B) Measure the maximum seismic-wave amplitude for each seismogram.

C) Convert your amplitude measurements into mm using the scaling information provided with each seismogram.

D) Plot lag time and amplitude on Figure 2.1 to determine magnitude.

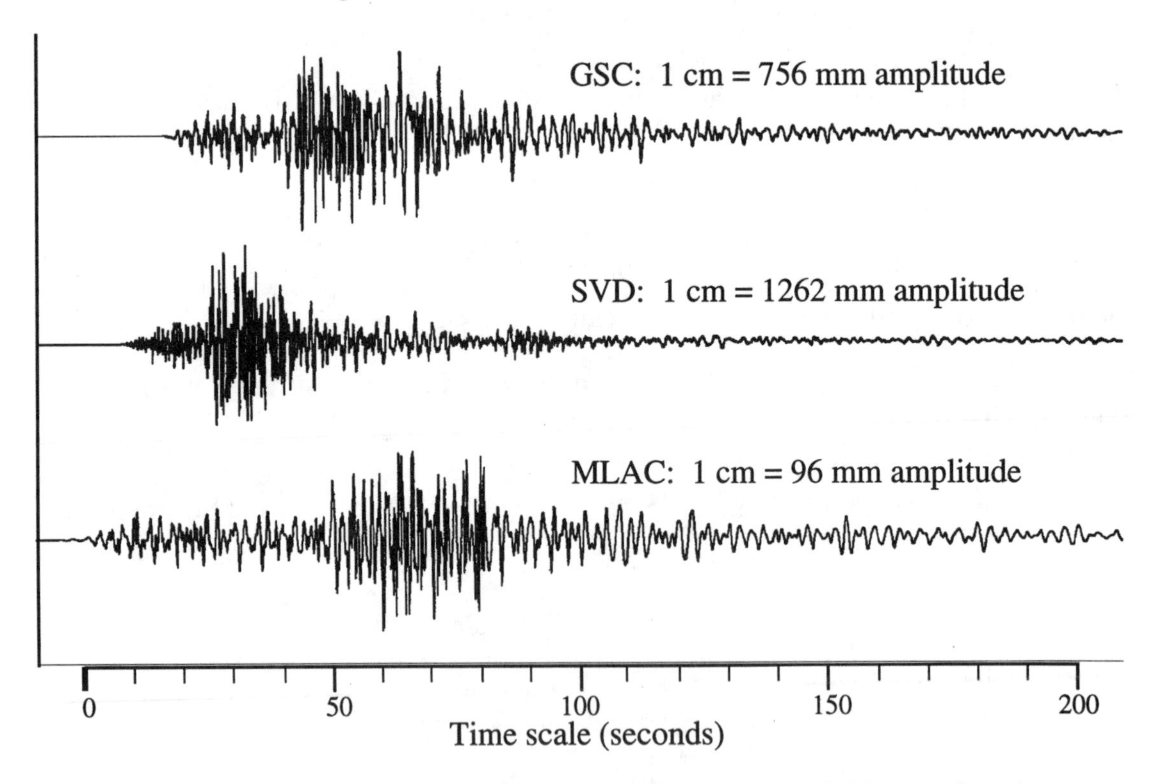

Figure 2.2. Three seismograms that record the 1994 Northridge earthquake.

7) Assume that the Moment magnitude (M_w) for the Northridge earthquake equals the Richter magnitude (M). What was the seismic moment (M_o) of that earthquake?

8) Aftershocks that followed the Northridge earthquake defined a fault plane about 15 km long, dipping about 40° to the south-southwest, with rupture from the focal depth (about 19 km) up to a depth of 8 km. Using this information (summarized in the diagram below), calculate the surface area (A) of the fault that caused the Northridge earthquake.

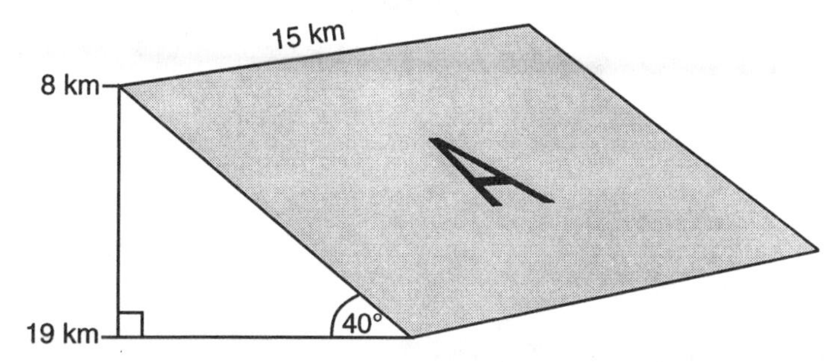

9) Using Equation 2.1 and your results from Questions 7 and 8, find the average fault displacement that occurred during the Northridge earthquake.

10) 1.0 dyne·cm is the energy necessary to lift a 1.0 g mass a distance of 0.001 cm. Assume that the Northridge earthquake uplifted an area of 500 km^2 by an average of 35 cm, that uplift occurred from 8 km depth to the surface, and that the average density of the uplifted crust is 3000 kg/m^3. What percentage of the earthquake's total energy (seismic moment) went into uplifting that area? Where does the rest of the energy go?

EARTHQUAKE INTENSITY

Earthquake intensity is defined as the strength of seismic shaking at a given location. Whereas an earthquake has just a single magnitude, it will have many different intensities at different locations. In general, areas closest to the epicenter experience the highest intensities, and shaking diminishes in strength farther away. This phenomenon is the result of seismic-wave *attenuation,* which is the reduction in wave amplitude and wave energy as they travel away from their source.

In order to study the patterns of earthquake intensity during different earthquakes, a system has been devised to assign specific numbers to different levels of shaking. The *Mercalli scale* was developed in 1902 and modified in the 1930s. The Mercalli scale assigns a numerical value, from Roman numeral I to XII, to the intensity of seismic shaking at any one particular location. The criteria for each Mercalli intensity are listed in Table 2.1. Figure 2.3 illustrates the distribution of intensities during an earthquake that struck southern Michigan in 1947. Note that the lines of equal intensity (called *isoseismal lines)* on Figure 2.3 are not perfect circles. The intensity of ground shaking can be influenced strongly by local and regional geology, by focusing of seismic waves, and by near-surface sediments. For example, *material amplification* (amplification of shaking by near-surface material) can cause some of the worst damage during earthquakes. During both the 1906 and the 1989 earthquakes that struck San Francisco, the worst shaking damage occurred in the city's Marina District, which is built on artificial fill added to San Francisco Bay. Jets of fluidized sediment during the 1989 earthquake unearthed debris from buildings destroyed in 1906.

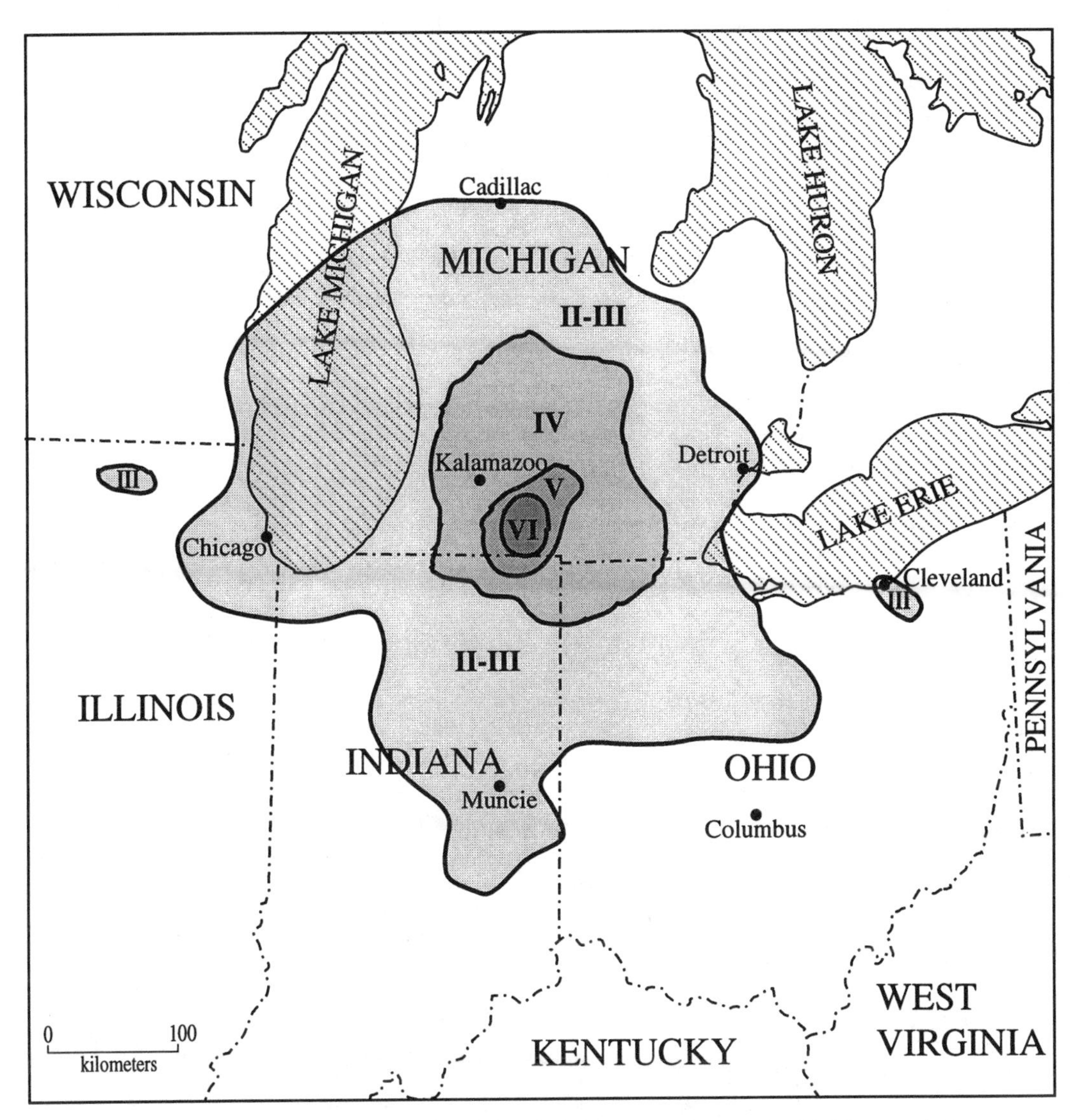

Figure 2.3. An earthquake that hit Michigan on Aug. 10, 1947 was felt across much of the Midwest. (After Stover and Coffman, 1993)

The descriptions of earthquake shaking used in the following exercise are summaries of the responses to 664 questionnaires mailed to post offices, police stations, and fire stations in the area affected by the 1994 Northridge earthquake (Dewey et al, 1995). You will use these summaries to assign a Mercalli intensity to each location (see Example 2.3).

Table 2.1. Modified Mercalli scale (Abridged; After Wood and Neumann, 1931)

I.	Not felt, except by a very few, under especially favorable circumstances.
II.	Felt only by a few persons at rest, especially on upper floors of buildings. Delicately suspended objects may swing.
III.	Felt quite noticeably indoors, especially on upper floors of buildings, but many people do not recognize it as an earthquake. Standing automobiles may rock slightly. Vibrations like a passing truck.
IV.	During the day, felt indoors by many, outdoors by few. At night, some awakened. Dishes, windows, doors disturbed; cracking sounds from walls. Sensation like heavy truck striking building. Standing automobiles rock noticeably.
V.	Felt by nearly everyone; many awakened. Some dishes, windows, etc. broken; a few instances of cracked plaster; unstable objects overturned. Disturbance of trees, poles, and other tall objects sometimes noticed. Pendulum clocks may stop.
VI.	Felt by all; many frightened and run outdoors. Some heavy furniture moved; a few instances of fallen plaster or damaged chimneys. Structural damage slight.
VII.	Everybody runs outdoors. Damage negligible in buildings of good design and construction; slight to moderate in well-built ordinary structures; considerable in poorly-built or badly-designed structures; some chimneys broken. Noticed by people driving.
VIII.	Damage slight in specially-engineered structures; considerable in ordinary structures. Many buildings with partial collapse. Panel walls thrown out of some frame structures. Fall of chimneys, factory stacks, columns, monuments, masonry walls. Heavy furniture overturned. Sand and mud ejected in small amounts. Changes in well water.
IX.	Damage considerable in all structures. Well-designed frame structures thrown out of plumb; partial collapse in many substantial buildings. Buildings shifted off foundations. Ground conspicuously cracked. Underground pipes broken.
X.	Some well-built wooden structures destroyed; most masonry and frame structures destroyed, including foundations. Ground badly cracked. Rails bent. Landslides considerable on river banks and other steep slopes. Shifted sand and mud.
XI.	Few, if any, masonry structures remain standing. Bridges destroyed. Broad fissures in ground. Underground pipe lines completely out of service. Earth slumps and land slips in soft ground. Rails bent greatly.
XII.	Damage total. Waves seen on ground surface. Lines of sight distorted. Objects thrown upward into the air.

Example 2.3.

Assign a Mercalli intensity for the earthquake effects summarized below:

> On the UCLA campus, Royce Hall was closed because of damage to its two masonry towers. A 15 by 75 ft section of heavy ceiling fell in a campus auditorium. About one-third of the books in the University Research Library fell, piling up three feet deep in places. Elsewhere in the vicinity, masonry fences were destroyed; underground pipes were put out of service, plaster walls sustained large cracks; and some chimneys were damaged or fell.

As you compare this description with Table 2.1, you'll see that some of the details in the description are directly pertinent to the criteria in the table, and others are not. For example, damage to chimneys is a criterion for Mercalli intensities VI, VII, and VIII. The description of the UCLA campus shows that some buildings were partially damaged and implies that most others sustained little or no structural damage. Mercalli intensity VII is closest, with its criterion, "Damage negligible in buildings of good design; slight to moderate in well-built ordinary structures." Other details are consistent with shaking that was greater than intensity VI but less than intensity VIII.

INTENSITY OF THE 1994 NORTHRIDGE EARTHQUAKE

The Northridge earthquake was felt over an area in excess of 214,000 km^2 (82,000 mi^2), from Ensenada, Mexico to Turlock in California's Central Valley to Richfield, Utah. As mentioned in Exercise 1, the earthquake killed 33 people and caused an estimated $20 billion in damage, making it at the time the costliest earthquake ever. In this exercise you will read a series of descriptions of seismic shaking and damage associated with the Northridge earthquake (from Dewey et al, 1995), assign Mercalli intensities to those descriptions, and map the isoseismal lines.

A) For each description listed on the following pages, assign a Mercalli intensity for that location.

B) When you have completed the list, plot the different intensities on Figure 2.4.

C) Assigning Mercalli intensities is not a purely objective process; you may wish to go back to some of those descriptions and reassess them after looking at the regional pattern.

D) The final step is to contour the intensities on Figure 2.4. Note that this is *not* a connect-the-dots puzzle. Isoseismal lines *enclose* all intensities equal to, or greater than, a given value.

E) Answer the questions that follow.

Acton ______: Many objects fell from store shelves; many homes sustained minor damage; pictures fell; a few windows cracked; small appliances moved.

Anaheim ______: A scoreboard structure at Anaheim stadium collapsed and damaged over 1000 seats. A few windows cracked; a few small objects overturned; a few people ran out of buildings. Felt by almost all people; many awakened. Hanging pictures swayed. Trees and bushes shook slightly to moderately.

Burbank ______: Police reported that the southwest section of town suffered the worst damage, including damage to the airport and the power plant. Masonry walls were destroyed and underground pipes broken. Many windows were broken out; light and heavy furniture was overturned. Many chimneys fell.

Chatsworth ______: Masonry fences were destroyed; underground pipes were put out of service; sidewalks and roadways sustained large cracks; many chimneys were broken at the roofline or fell. A beverage-can plant sustained extensive damage to manufacturing equipment and had to cease operations for an estimated ten weeks. One person was killed by falling objects in his house.

Compton ______: Plaster walls sustained hairline cracks and separated from ceiling or floor; a few chimneys were cracked; several small items were overturned; several items fell from store shelves; water splashed onto sides of swimming pools.

Downtown Los Angeles ______: The roof partially collapsed in the City Hall parking structure, and water lines broke in City Hall and in Parker Center. Interior and exterior walls sustained large cracks; some windows were broken out; a few small objects overturned and fell.

Fillmore ______: The picturesque old part of town, which contained many brick buildings built early this century, suffered extensive damage; some wood-frame houses were shifted off their foundations. Natural gas from a damaged pipeline ignited, and the fire spread to a nearby mobile home park.

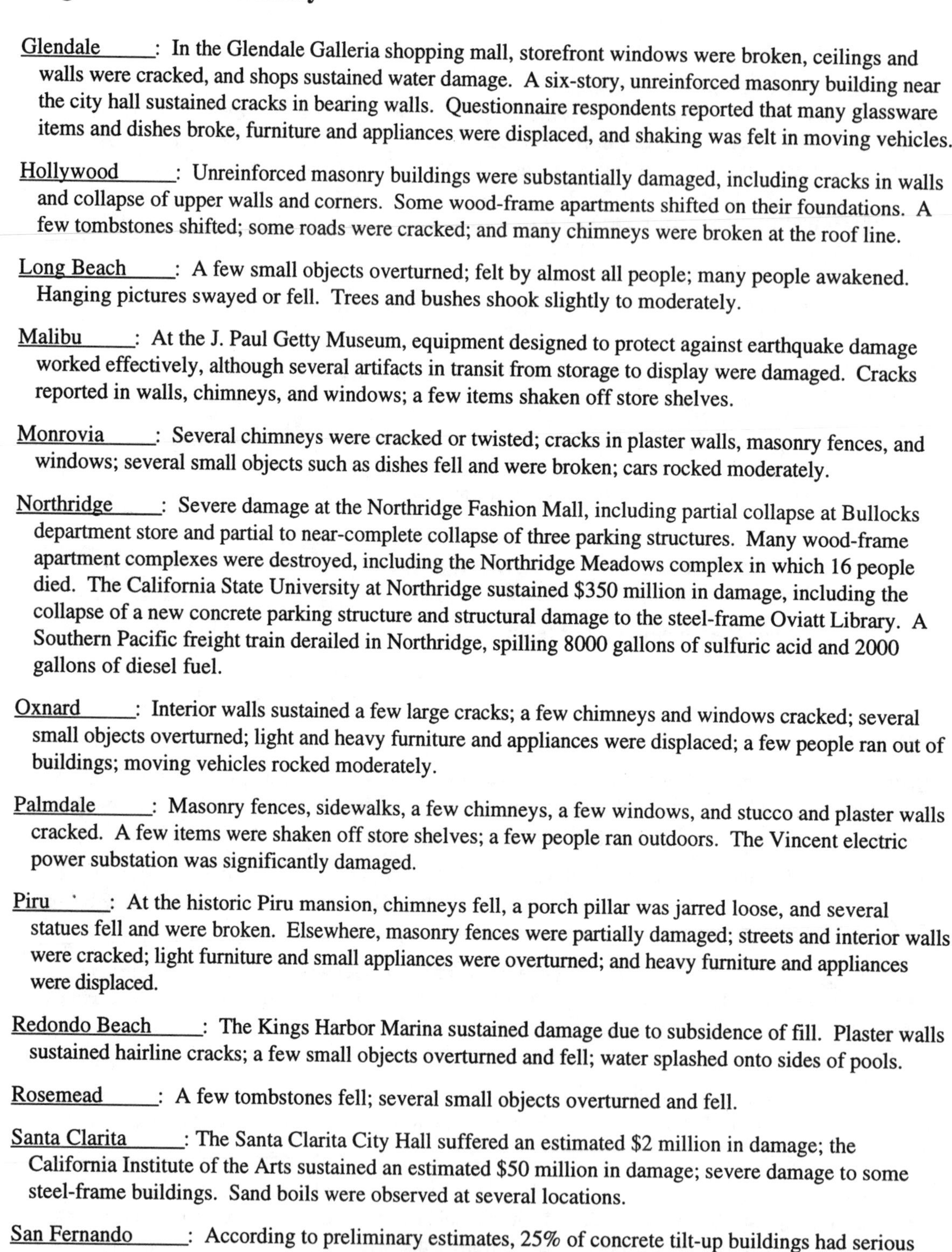

Magnitude and Intensity

Glendale: In the Glendale Galleria shopping mall, storefront windows were broken, ceilings and walls were cracked, and shops sustained water damage. A six-story, unreinforced masonry building near the city hall sustained cracks in bearing walls. Questionnaire respondents reported that many glassware items and dishes broke, furniture and appliances were displaced, and shaking was felt in moving vehicles.

Hollywood: Unreinforced masonry buildings were substantially damaged, including cracks in walls and collapse of upper walls and corners. Some wood-frame apartments shifted on their foundations. A few tombstones shifted; some roads were cracked; and many chimneys were broken at the roof line.

Long Beach: A few small objects overturned; felt by almost all people; many people awakened. Hanging pictures swayed or fell. Trees and bushes shook slightly to moderately.

Malibu: At the J. Paul Getty Museum, equipment designed to protect against earthquake damage worked effectively, although several artifacts in transit from storage to display were damaged. Cracks reported in walls, chimneys, and windows; a few items shaken off store shelves.

Monrovia: Several chimneys were cracked or twisted; cracks in plaster walls, masonry fences, and windows; several small objects such as dishes fell and were broken; cars rocked moderately.

Northridge: Severe damage at the Northridge Fashion Mall, including partial collapse at Bullocks department store and partial to near-complete collapse of three parking structures. Many wood-frame apartment complexes were destroyed, including the Northridge Meadows complex in which 16 people died. The California State University at Northridge sustained $350 million in damage, including the collapse of a new concrete parking structure and structural damage to the steel-frame Oviatt Library. A Southern Pacific freight train derailed in Northridge, spilling 8000 gallons of sulfuric acid and 2000 gallons of diesel fuel.

Oxnard: Interior walls sustained a few large cracks; a few chimneys and windows cracked; several small objects overturned; light and heavy furniture and appliances were displaced; a few people ran out of buildings; moving vehicles rocked moderately.

Palmdale: Masonry fences, sidewalks, a few chimneys, a few windows, and stucco and plaster walls cracked. A few items were shaken off store shelves; a few people ran outdoors. The Vincent electric power substation was significantly damaged.

Piru: At the historic Piru mansion, chimneys fell, a porch pillar was jarred loose, and several statues fell and were broken. Elsewhere, masonry fences were partially damaged; streets and interior walls were cracked; light furniture and small appliances were overturned; and heavy furniture and appliances were displaced.

Redondo Beach: The Kings Harbor Marina sustained damage due to subsidence of fill. Plaster walls sustained hairline cracks; a few small objects overturned and fell; water splashed onto sides of pools.

Rosemead: A few tombstones fell; several small objects overturned and fell.

Santa Clarita: The Santa Clarita City Hall suffered an estimated $2 million in damage; the California Institute of the Arts sustained an estimated $50 million in damage; severe damage to some steel-frame buildings. Sand boils were observed at several locations.

San Fernando: According to preliminary estimates, 25% of concrete tilt-up buildings had serious structural damage, including partial collapse. Minor ground cracking occurred. A reinforced-masonry building under construction sustained significant damage.

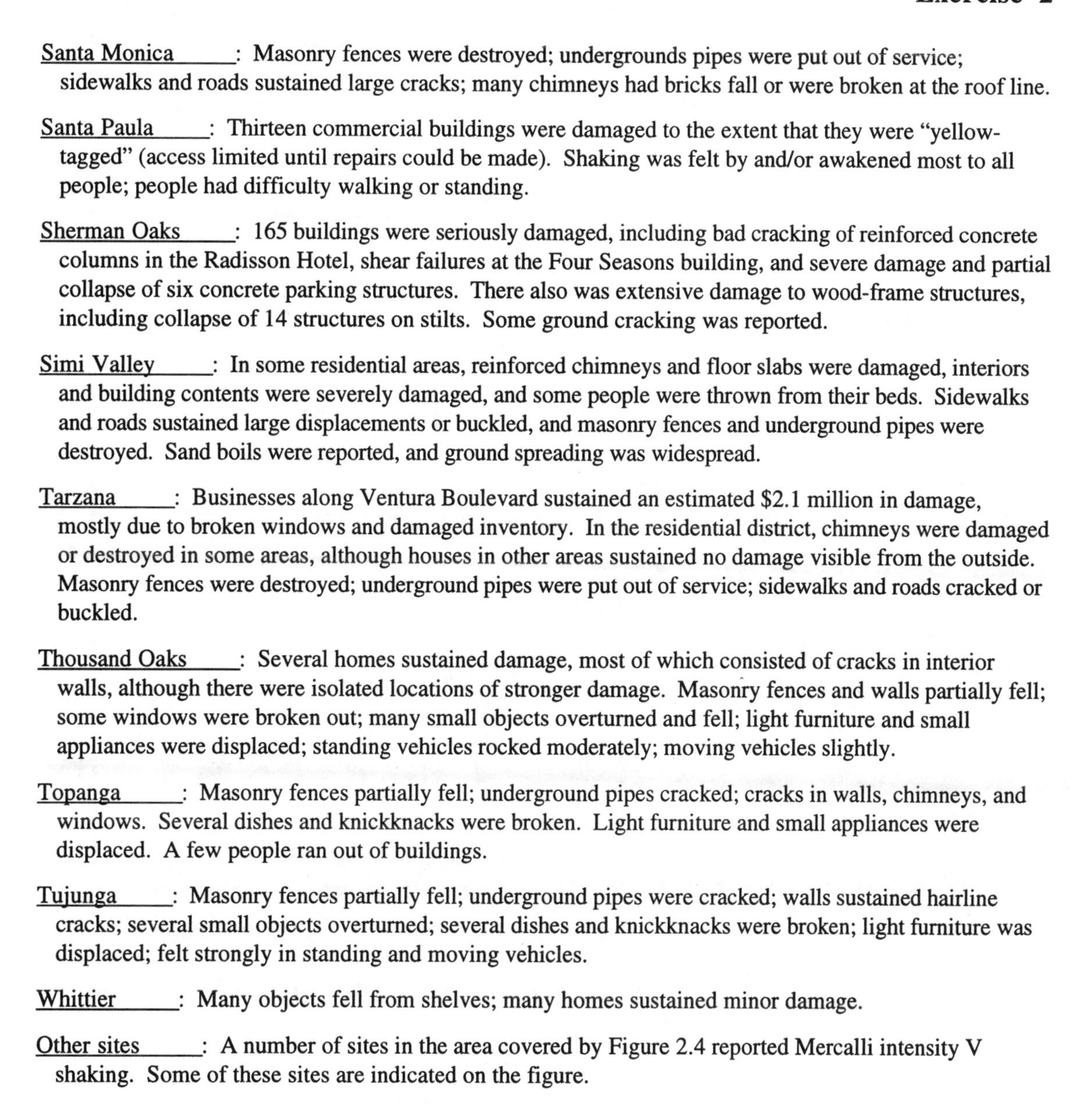

Santa Monica ____: Masonry fences were destroyed; undergrounds pipes were put out of service; sidewalks and roads sustained large cracks; many chimneys had bricks fall or were broken at the roof line.

Santa Paula ____: Thirteen commercial buildings were damaged to the extent that they were "yellow-tagged" (access limited until repairs could be made). Shaking was felt by and/or awakened most to all people; people had difficulty walking or standing.

Sherman Oaks ____: 165 buildings were seriously damaged, including bad cracking of reinforced concrete columns in the Radisson Hotel, shear failures at the Four Seasons building, and severe damage and partial collapse of six concrete parking structures. There also was extensive damage to wood-frame structures, including collapse of 14 structures on stilts. Some ground cracking was reported.

Simi Valley ____: In some residential areas, reinforced chimneys and floor slabs were damaged, interiors and building contents were severely damaged, and some people were thrown from their beds. Sidewalks and roads sustained large displacements or buckled, and masonry fences and underground pipes were destroyed. Sand boils were reported, and ground spreading was widespread.

Tarzana ____: Businesses along Ventura Boulevard sustained an estimated $2.1 million in damage, mostly due to broken windows and damaged inventory. In the residential district, chimneys were damaged or destroyed in some areas, although houses in other areas sustained no damage visible from the outside. Masonry fences were destroyed; underground pipes were put out of service; sidewalks and roads cracked or buckled.

Thousand Oaks ____: Several homes sustained damage, most of which consisted of cracks in interior walls, although there were isolated locations of stronger damage. Masonry fences and walls partially fell; some windows were broken out; many small objects overturned and fell; light furniture and small appliances were displaced; standing vehicles rocked moderately; moving vehicles slightly.

Topanga ____: Masonry fences partially fell; underground pipes cracked; cracks in walls, chimneys, and windows. Several dishes and knickknacks were broken. Light furniture and small appliances were displaced. A few people ran out of buildings.

Tujunga ____: Masonry fences partially fell; underground pipes were cracked; walls sustained hairline cracks; several small objects overturned; several dishes and knickknacks were broken; light furniture was displaced; felt strongly in standing and moving vehicles.

Whittier ____: Many objects fell from shelves; many homes sustained minor damage.

Other sites ____: A number of sites in the area covered by Figure 2.4 reported Mercalli intensity V shaking. Some of these sites are indicated on the figure.

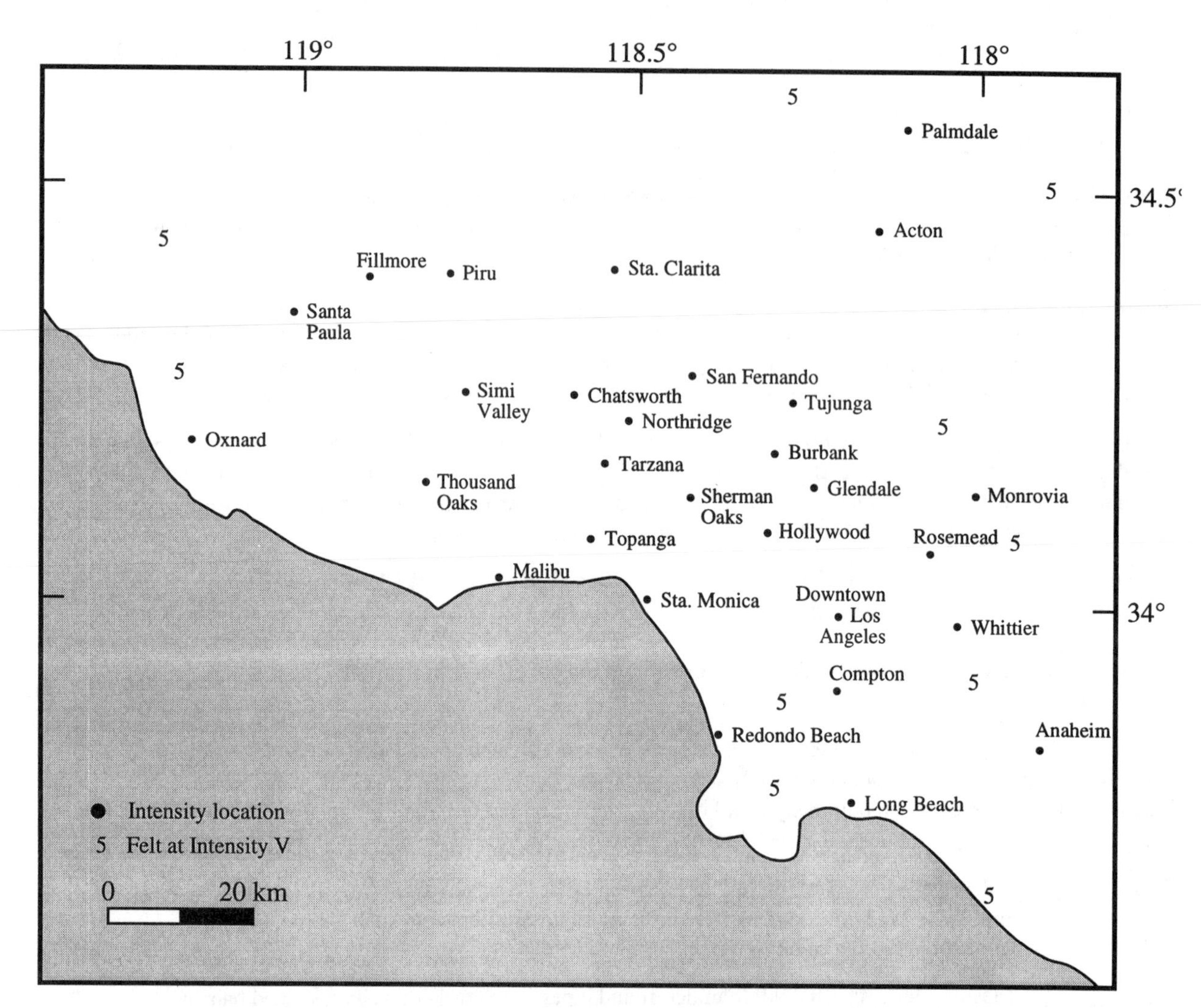

Figure 2.4. Location map for the greater Los Angeles metropolitan area.

11) Are the isoseismic lines on the map perfect circles? If not, why not?

12) The maximum intensity reported for the Northridge earthquake by the U.S. Geological survey was IX. Intensity summaries like the ones in this exercise (as well as television and newspaper reports) focus on destruction and deemphasize buildings or areas that were not damaged. In what way would this lead you to overestimate Mercalli intensities?

BIBLIOGRAPHY

Bolt, B.A., 1978. Earthquakes. W.H. Freeman: San Francisco.

Bolt, B.A., 1993. Earthquakes (3rd Edition). W.H. Freeman: San Francisco.

Bruhn, J.N., 1968. Seismic moment, seismicity, and rate of slip along major fault zones. Journal of Geophysical Research, 73: 777-784.

Dewey, J.W., B.G. Reagor, L. Dengler, and K. Moley, 1995. Intensity distribution and isoseismal maps for the Northridge, California earthquake of January 17, 1994. U.S. Geological Survey Open-File Report 95-92.

Hanks, T.C., and H. Kanamori, 1979. A moment magnitude scale. Journal of Geophysical Research, 84: 2348-2350.

Pinter, N., 1995. Faulting on the Volcanic Tableland, Owens Valley, California. Journal of Geology, 103: 73-83.

Stover, C.W., and J.L. Coffman, 1993. Seismicity of the United States, 1568-1989 (Revised): U.S. Geological Survey Professional Paper 1527.

Wood, H.O., and F. Neumann, 1931. Modified Mercalli Intensity scale of 1931. Seismological Society of America Bulletin, 21: 277-283.

Acknowledgements: The author would like to acknowledge the assistance of James Dewey of the U.S. Geological Survey in Denver. Most of the intensity data for the Northridge earthquake came from Dewey et al, 1995. Seismograms for magnitude determination were assembled and prepared by Paul Roberts of the California Institute of Technology. My thanks to these scientists for letting me utilize their results.

EXERCISE 3

TOPOGRAPHIC MAPS AND AIR PHOTOGRAPHS

Supplies Needed

- calculator
- metric ruler
- red pen or pencil
- piece of graph paper
- pocket stereoscope
- sheet of clear acetate

PURPOSE

Active tectonics is the study of earthquakes, faulting, and crustal deformation that are occurring, or have occurred in the recent geologic past. Earthquakes are perhaps the most dramatic of these processes, but geologists can only study earthquakes that occurred in historical time, which is a period that ranges from a few decades up to two or three thousand years in different parts of the world. The Earth's landscape, however, is shaped over periods of thousands of years up to a few million years. As a result, surface landforms in active tectonic settings can record hundreds or even thousands of earthquakes and faulting events. Topographic maps and air photographs are among the principal tools for investigating surface features. This exercise serves as a review of the basics of topographic-map and air-photo use, and as an introduction to the identification and analysis of tectonic landscapes.

INTRODUCTION TO TECTONIC LANDSCAPES

Earthquakes and faulting leave their marks on the landscape. Over periods of thousands to millions of years, tectonic processes shape the surface of the Earth. This branch of geology is called *tectonic geomorphology,* and it is a major element of the broader field of *active tectonics.*

Different landscapes commonly are classified according to the predominant process or processes that shape the surface. Figure 3.1 lists a few of the characteristic landforms associated with the major *morphogenetic systems* (landscape-shaping systems). In some locations, tectonic processes predominate, but it is more common that tectonic processes (and indeed all geomorphic processes) operate in combination with other activity. Several of the exercises later in this book focus on specific systems, while others outline techniques useful in a variety of different settings.

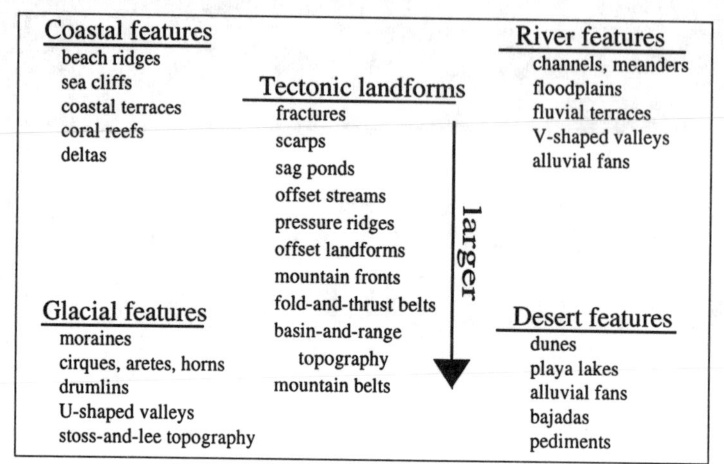

Figure 3.1. Examples of landforms in a few of the Earth's morphogenetic systems.

In this exercise, you will get some panoramic views of landscapes that dramatically illustrate tectonic processes at work. Exercise 1 reviewed the three basic types of faults (normal, reverse, and strike-slip). Active faults of each of these three types are associated with distinct assemblages of landforms. Figure 3.2 illustrates some of the landforms typical of active strike-slip faults. We investigate tectonic landforms for two main reasons: 1) to identify the pattern, character, and rates of active tectonic processes, and 2) to understand how these processes shape the Earth's landscape. This chapter provides you with some of the fundamental tools to carry out those two objectives.

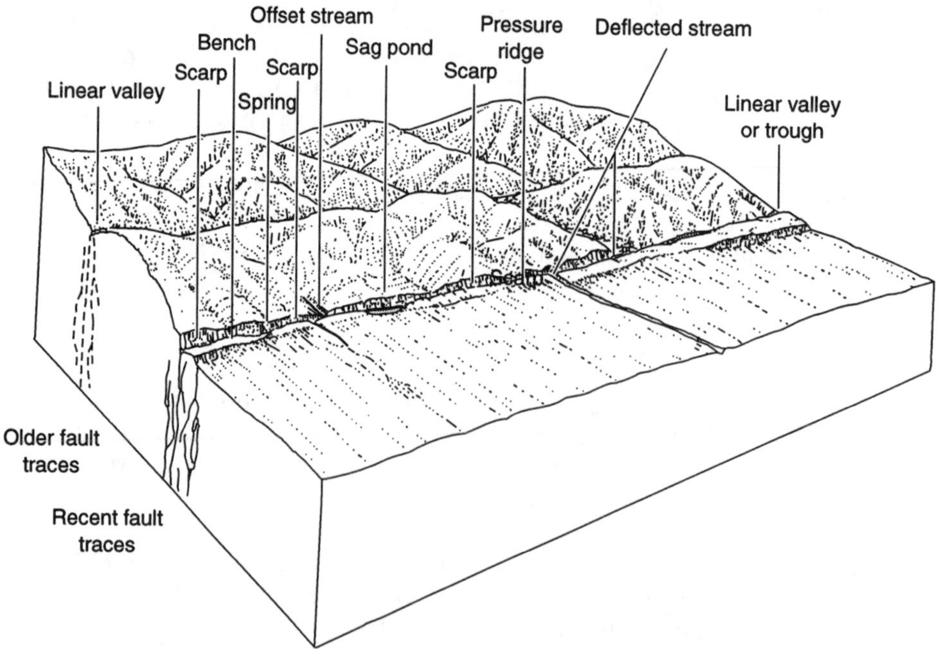

Figure 3.2. Landforms associated with active strike-slip faulting. (After Borcherdt, 1975)

TOPOGRAPHIC MAPS

All maps are graphical representations of geographical information. That information can include types of rock found at or near the surface (geologic maps), surface landforms and deposits (geomorphic maps), rainfall (precipitation and climate maps), locations and names of streets (road maps), and almost any other variety of spatial data. Topographic maps commonly include a range of information:

- elevation data (contour lines)
- cultural data (roads, buildings, power lines, etc.)
- geographical framework (streams, shorelines, and vegetation cover)

Of the three types of information listed above, elevation data is the most useful to geologists. In fact, topographic ("topo") maps are one of the primary tools of field-based geologists for locating themselves and geologic features, for assessing and measuring landforms, and as bases for constructing geologic and geomorphic maps.

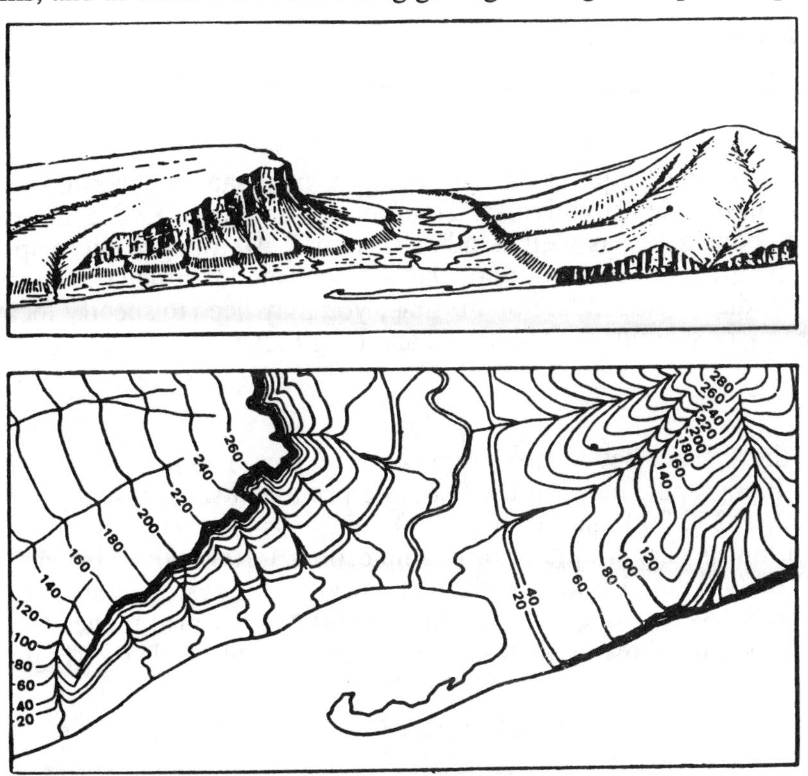

Figure 3.3. An example of how three-dimensional topography is shown using contour lines. (From U.S. Geological Survey)

Elevation is shown on a topographic map in the form of *contour lines.* A contour line is a line of equal elevation across a map. By definition, a contour line never climbs up or down and never crosses another contour line. Figure 3.3 illustrates how a landscape is depicted as a series of contour lines. Note that the contour lines are close together where slopes are steep, and the lines are farther apart where slopes are less steep. Remember the following rules whenever you look at a topo map or use contour lines for elevation control:

- contour lines are defined as lines of equal elevation
- contour lines may join, but *only* at a vertical cliff
- contour lines are deflected up valleys, forming upslope-pointing "V"s where they cross streams (see Figure 3.3)
- *contour interval* is defined as the vertical distance between adjacent contour lines
- contour lines are only approximations of elevation. The rule of thumb is that the accuracy of any one contour line is:

  +/- 0.5 * (one contour interval).

In addition to elevation, you will utilize three simple types of information from topographic maps:

- location
- distance
- slope

Location:

The physical position of a point on a topographic map can be specified in several ways: 1) latitude and longitude, 2) Universal Transverse Mercator (UTM) coordinates, 3) township and range coordinates, as well as other systems. All topographic maps in this book show latitude (degrees north or south of the equator) and longitude (degrees east or west of the prime meridian). Depending on the map, you may need to specify locations to degrees (°), minutes ('), or seconds (") of latitude and longitude (1° = 60'; 1' = 60").

Distance:

You can find the distance between any two points on a map so long as you know the *scale* of that map. Topographic maps in the U.S. are printed at several scales, including 1:24,000, 1:100,000, 1:250,000, and 1:1,000,000. Expressing a map's scale as a ratio makes calculating distances very simple. For example, on a 1:100,000 scale map, 1.0 cm on the map equals 100,000 cm (1.0 km) in the real world. This system works regardless of the units one chooses – 1.0 map inch equals 100,000 real-world inches (about 2.5 miles). The other method for showing map scale is using a graphic scale bar (Figure 3.4).

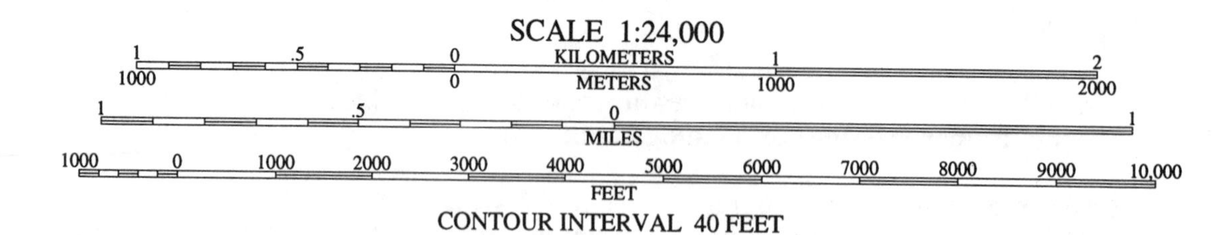

Figure 3.4. Typical scale bar from a topographic map, showing scale for both English and metric units. (From U.S. Geological Survey)

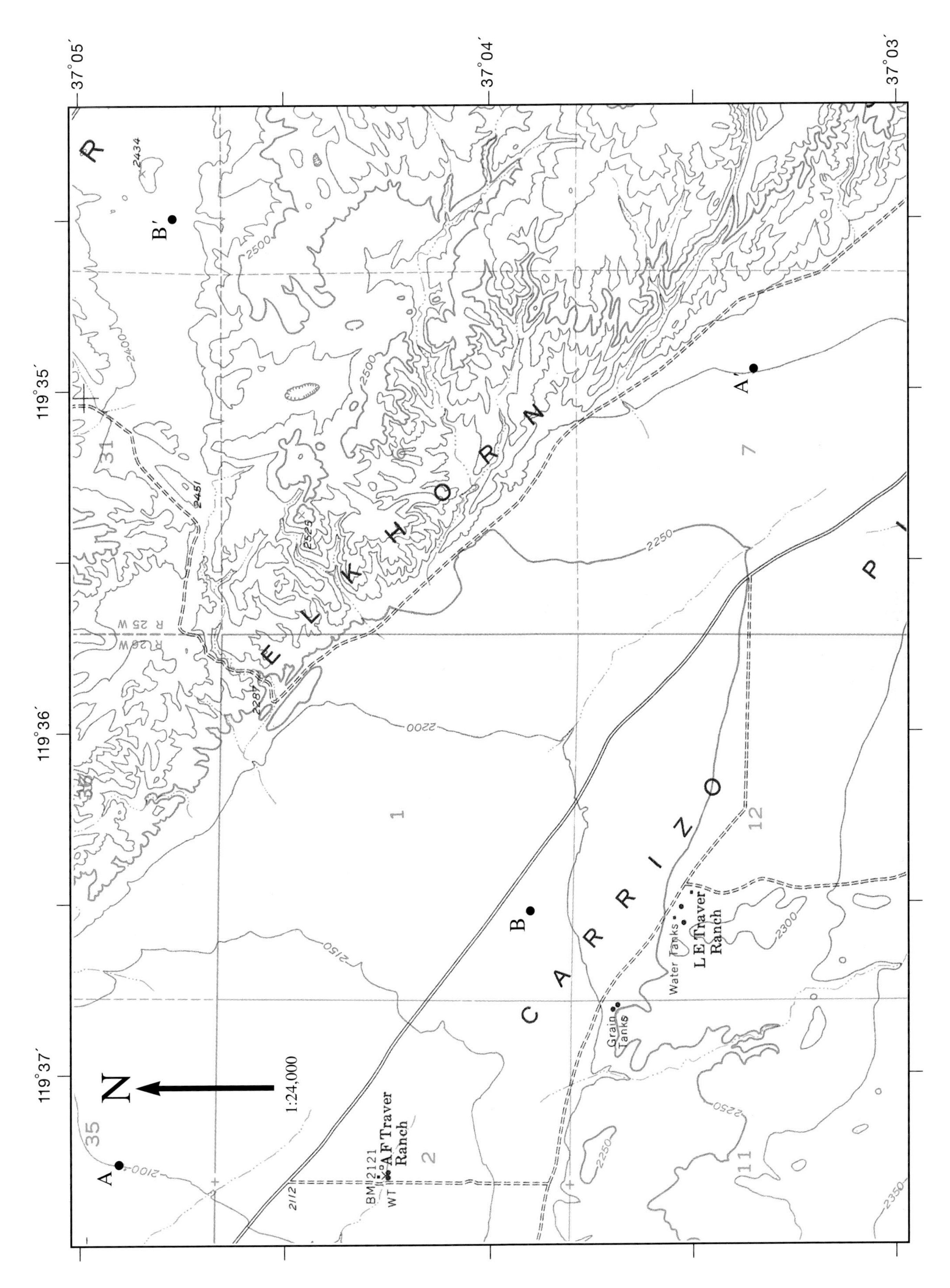

Figure 3.5 Portion of topographical map of the Elkhorn Hills. Scale: 1:24,000.

Slope:

As stated earlier, steep slopes are shown on a topographic map by closely-spaced contour lines. Where contour lines are farther apart, the topography is less steep. The average steepness between any two points can be expressed either in terms of *slope gradient* or *slope angle.* The expression for calculating slope gradient is:

$$\text{slope gradient} = \frac{\text{rise}}{\text{run}} = \frac{\Delta z}{\Delta x} \qquad (3.1)$$

slope angle is simply the angle, the tangent of which is slope gradient:

$$\text{slope angle} = \arctan(\text{slope gradient}) \qquad (3.2)$$

Figure 3.5 is a portion of the Elkhorn Hills Quadrangle, a 1:24,000 scale topographic map that shows a portion of the San Andreas fault in California. Answer the following questions by using that map.

1) Give the latitude and longitude coordinates (to the nearest 10 seconds) of Point A.

2) Calculate the real-world distance between Point A and Point A'.

3) Calculate the slope gradient and slope angle between Point A and Point A'.

4) In a red pen or colored pencil, trace the San Andreas fault on Figure 3.5. What fault-related featured can you make out on this map?

Topographic Profiles:

Topographic profiles are vertical cross-sections that show changes in the elevation of the surface along a single line. You will use topographic profiles several times in this book, and they are powerful tools in many areas of geology. Figure 3.6 illustrates one topographic profile constructed from the simplified map shown. Notice that the profile is a smooth curve that interpolates between the points where contour lines cross the profile line. For example, the highest contour line near the crest of the ridge is 270 m, but the actual crest of the ridge must be higher than 270 m and lower than 280 m. Also note that the profile has vertical exaggeration – the horizontal axis uses a different scale than the vertical axis. Most topography on Earth is surprisingly flat when plotted with *no* vertical exaggeration. One of the powerful tools available when constructing topographic profiles is the ability to manipulate the exaggeration to best illustrate the topography.

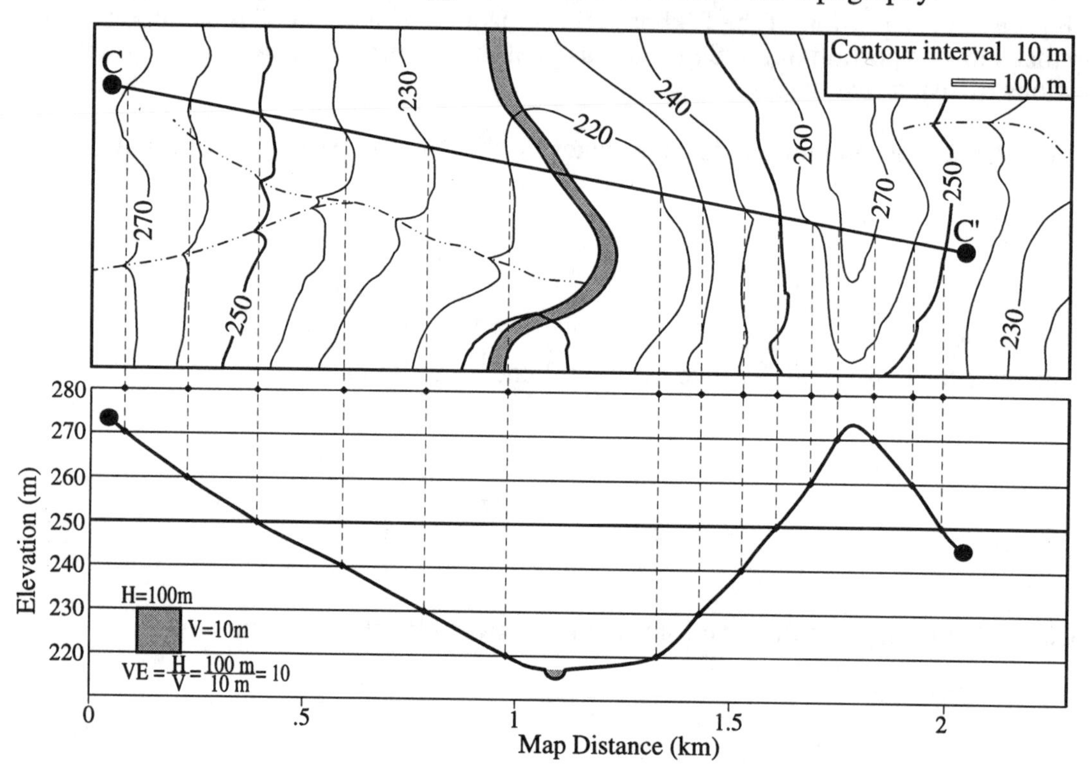

Figure 3.6. A simplified topographic map and a topographic profile between Points C and C' constructed from that map.

To construct a topographic profile between two points on a topographic map, follow these steps:

A) Place the *edge* of a piece of graph paper along the line connecting the starting point and the ending point of your profile (C and C' in Figure 3.6).

B) Along the edge of the graph paper, mark every point where the profile line intersects a contour line on the map, and label the elevation of that contour line. For profile lines that cross many contours, label only every *fifth* contour elevation.

C) Draw the axes of your profile on the graph paper. The horizontal (distance) axis has the same scale as the topographic map (for example, 1:24,000). Select a vertical (elevation) scale so that all of the elevations on your profile will fit onto the page. As on all graphs, label both axes appropriately.

D) For every mark at the edge of the graph paper, move parallel to the vertical axis and plot a point on the profile with the appropriate elevation.

E) Connect those points with a smooth line.

F) Calculate the vertical exaggeration factor (VE) for that profile (see Figure 3.6), where:

$$VE = \frac{H}{V} = \frac{\text{horizontal side of a square on the profile}}{\text{vertical side of a square on the profile}} \qquad (3.3)$$

5) Construct a topographic profile from Point B to Point B' on Figure 3.5. Just looking at the profile, can you identify where the San Andreas fault crosses the profile line? If you can, mark the profile at that point with a red arrow.

AIR PHOTOGRAPHS

After topographic maps, air photographs are one of the most powerful tool a student of active tectonics has for quickly evaluating evidence of faulting or surface deformation in a region. Air photos are most effective in desert regions with few inhabitants. Forested regions obscure many small-scale landscape features, and urbanization tends to erase them. Although geologists sometimes would like to remove vegetation to get a better look at the rocks underneath, widespread deforestation is rarely feasible. Urbanization, however, can effectively be removed in some areas by obtained air photos that pre-date some or all of the growth. In the Los Angeles region, for example, a complete set of detailed vertical air photos was taken in 1929, before the most massive remodeling of the landscape occurred.

It should be mentioned without further delay that air photographs – including black-and-white, color, infrared, vertical, and oblique – are just one class of overhead images that are used in geology and many other applications. The surface of the Earth has been recorded at a variety of scales from aircraft and spacecraft using a variety of different instruments. The variety is so great, in fact, that there isn't room enough in this book to do the subject justice.

Air photographs, like topographic maps, are available at many different scales. *Large-scale* photos show a small area in great detail, while *small-scale* photos cover large areas with limited resolution. The scale of a particular photograph depends on the height of the aircraft from which it was taken and the focal length of the camera used. Air-photograph scale typically is given as a ratio, for example 1:12,000.

Another powerful tool presented by air photographs is the potential for *stereoscopic viewing*. Just as we human beings sense the three-dimensional details of the world around us using two eyes, a three-dimensional view of the Earth's surface can be obtained using two air photographs that show the same area, but from slightly different angles. Aircraft typically shoot many photos along a series of *flight lines* so that each photograph overlaps by about 60% with each adjacent photo. The three-dimensional view is available wherever two photographs overlap (Figure 3.7).

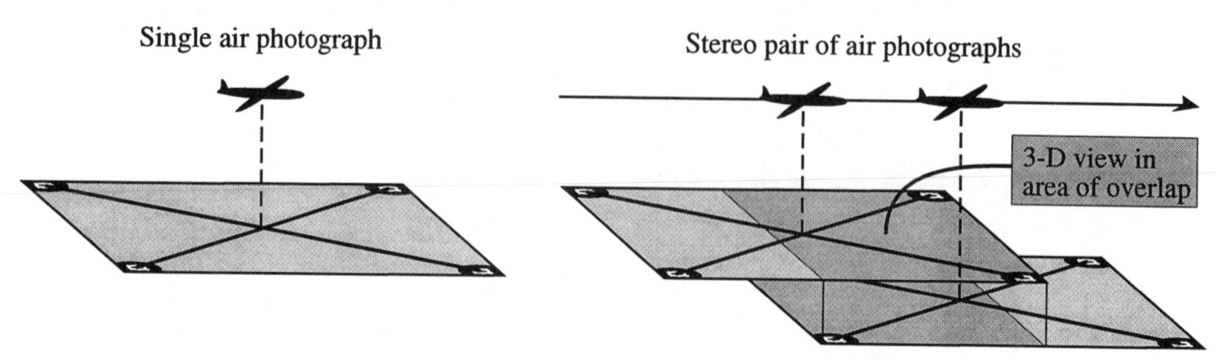

Figure 3.7. The difference between single and stereo air photographs.

You can read about stereo air viewing all day, but there is no substitute for actually using the photographs. Figure 3.8 is a stereo pair of 1:24,000-scale infrared photographs from Santa Cruz Island, California, illustrating landforms offset by the Santa Cruz Island fault. The two photographs are aligned for use with a pocket stereoscope. Position the stereoscope so that the right photograph is beneath one eyepiece, and the corresponding area of the left photo is beneath the other eyepiece. Move the stereoscope around until the two images appear to overlap, and then wait for the result.

6) What are the clumps of dark red on these photographs? Speculate why they are that color.

7) Can you get a three-dimensional view of the left-most side of Figure 3.8 (the left side of the left photo)? Why or why not?

8) The large stream valleys across the Santa Cruz Island fault in the area of these photographs are offset by as much as 300 m. The fault can be identified as a line of such offsets. Mark the location and trend of the fault on Figure 3.8 (if you don't want to permanently mark the photos, cover them with a sheet of clear acetate and use a felt-tip marker. Judging by the sense of offset, what type of fault is this?

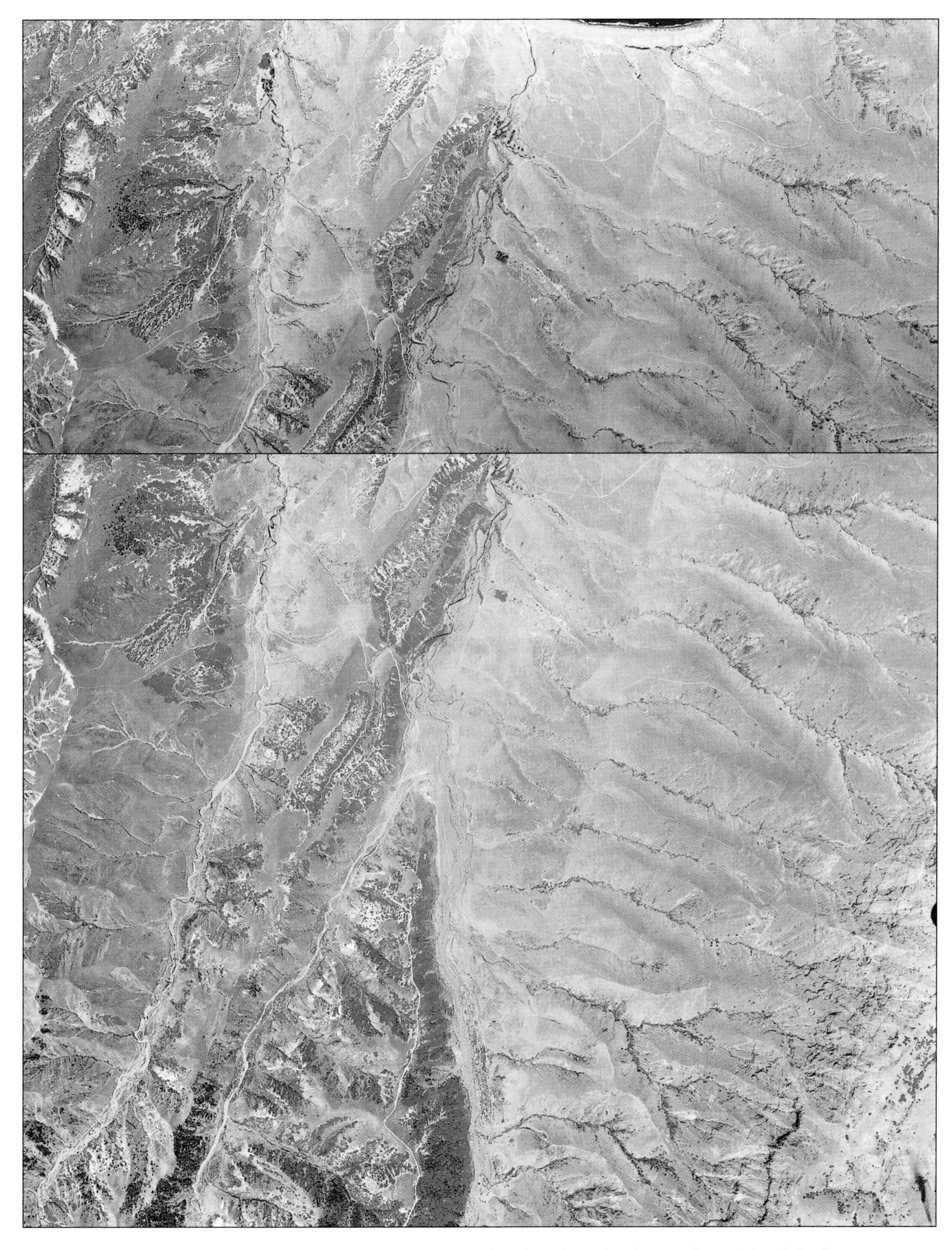

Figure 3.8 Stereo pair of color-infrared photographs showing the Santa Cruz Island fault, California. Scale: 1:24,000. (Images courtesy of Pacific Western of Santa Barbara, CA.)

CASE STUDY: TECTONIC DEFORMATION IN THE OWENS VALLEY, CALIFORNIA

The Owens Valley (Figure 3.9) is the westernmost basin of the Basin and Range province. The valley is a deep structural depression between the Sierra Nevada on the west and the White and Inyo mountain ranges on the east. The earliest geologists to visit this area were struck by its rugged topography; J.D. Whitney visited the valley in 1872:

> *"Both sides of the valley are bordered by extremely steep and elevated mountains, by which it is closed in, as if by two gigantic walls. On the west is the Sierra Nevada, with no pass across it of less than 12,000 feet [~3700 meters] in elevation; the crest of the range broken into a thousand pinnacles and battlements.... The Inyo and White Mountain range ... is one narrow crest, almost unique in its narrowness and steepness.... Dark, sombre, destitute of trees and water, and rarely whitened with snow, even for a short period, it presents a most striking contrast to the Sierra side of the valley."*
>
> (Whitney, 1872)

Like all of the basins of the Basin and Range, the Owens Valley was formed by normal faulting and regional stretching of western North America that occurred during the last several millions of years. Parts of the Basin and Range have more than doubled during this stretching.

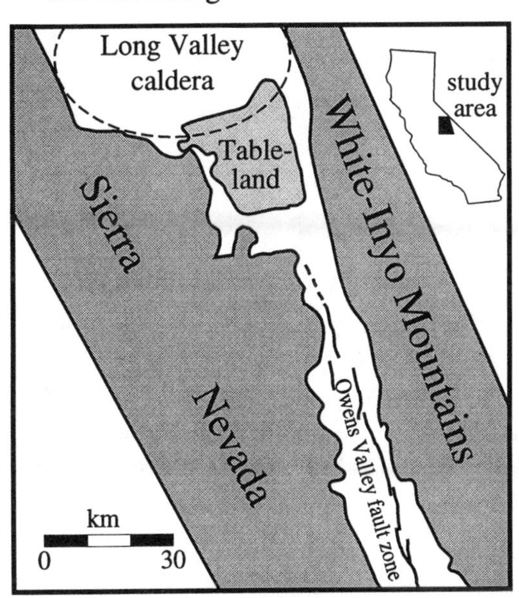

Figure 3.9. The Owens Valley.

Virtually every type of tectonic activity can be found in the Owens Valley: volcanic, hydrothermal, seismic, and surface deformation. For example, one of the most dramatic volcanic events on Earth occurred about 760,000 years ago, when Long Valley caldera (see Figure 3.9) erupted catastrophically, forming a huge volcanic crater and sending an estimated 600 km^3 of glowing ash rumbling southward. That ash now forms a deposit up to 1000 m thick called the "Volcanic Tableland." Continuing faulting and earthquake activity in the Owens Valley was dramatically manifested in 1872, when 110 km of the Owens Valley fault zone ruptured in a magnitude 7.8-8.0 earthquake. Tectonic processes combine with surface-sculpting by wind, water, and glacial ice to make a richly varied landscape.

Figure 3.10 contains parts of three adjacent air photographs from a flight line over the southern margin of the Volcanic Tableland. These three photos form two stereo pairs, providing three-dimensional coverage over two-thirds of the figure.

9) What are the long, north-south trending features that you can see on the Tableland? Tape a piece of acetate over the photos and, without using your stereoscope yet, trace as many of these features as you can identify with a felt-tip pen. Also trace the southern edge of the Volcanic Tableland.

10) Now use your pocket stereoscope to get the three-dimensional view of the Tableland. Are there additional north-south trending features that were not visible without the stereoscope? Trace these additional features. Speculate why they were not visible before. (Hint: You should be able to say whether these photos were taken in the morning or afternoon.)

11) South of the Volcanic Tableland, the Owens River has cut a broad floodplain. Above the active floodplain, several stream terraces record old positions of the river (Figure 3.11). Using the southern pair of air photographs, map the edges of the terraces on the acetate. Are the linear features that you mapped in Questions 9 and 10 also visible on the river terraces? Is the density of these features greater, less, or equal to their density on the Volcanic Tableland?

12) The T2 terrace formed about 10,000 years ago; the T3 terrace formed about 60,000 years ago (Pinter et al, 1994). The T4 terrace, which formed between 131,000 and 218,000 years ago, is not visible on Figure 3.10. What can you say about the age of faulting in the northern Owens Valley?

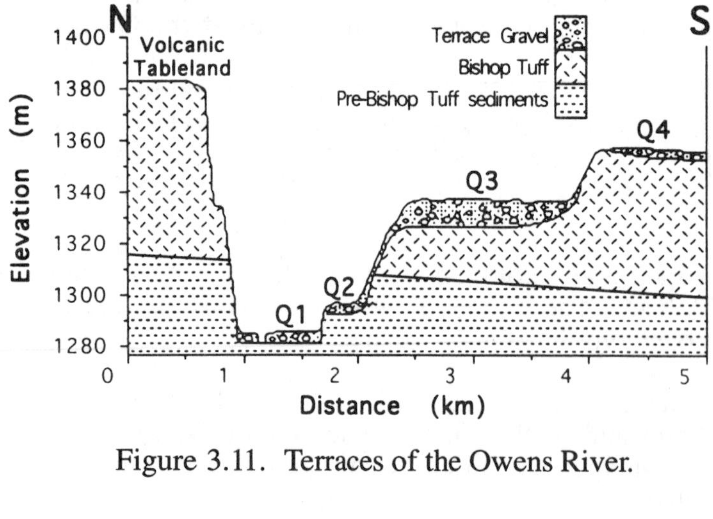

Figure 3.11. Terraces of the Owens River.

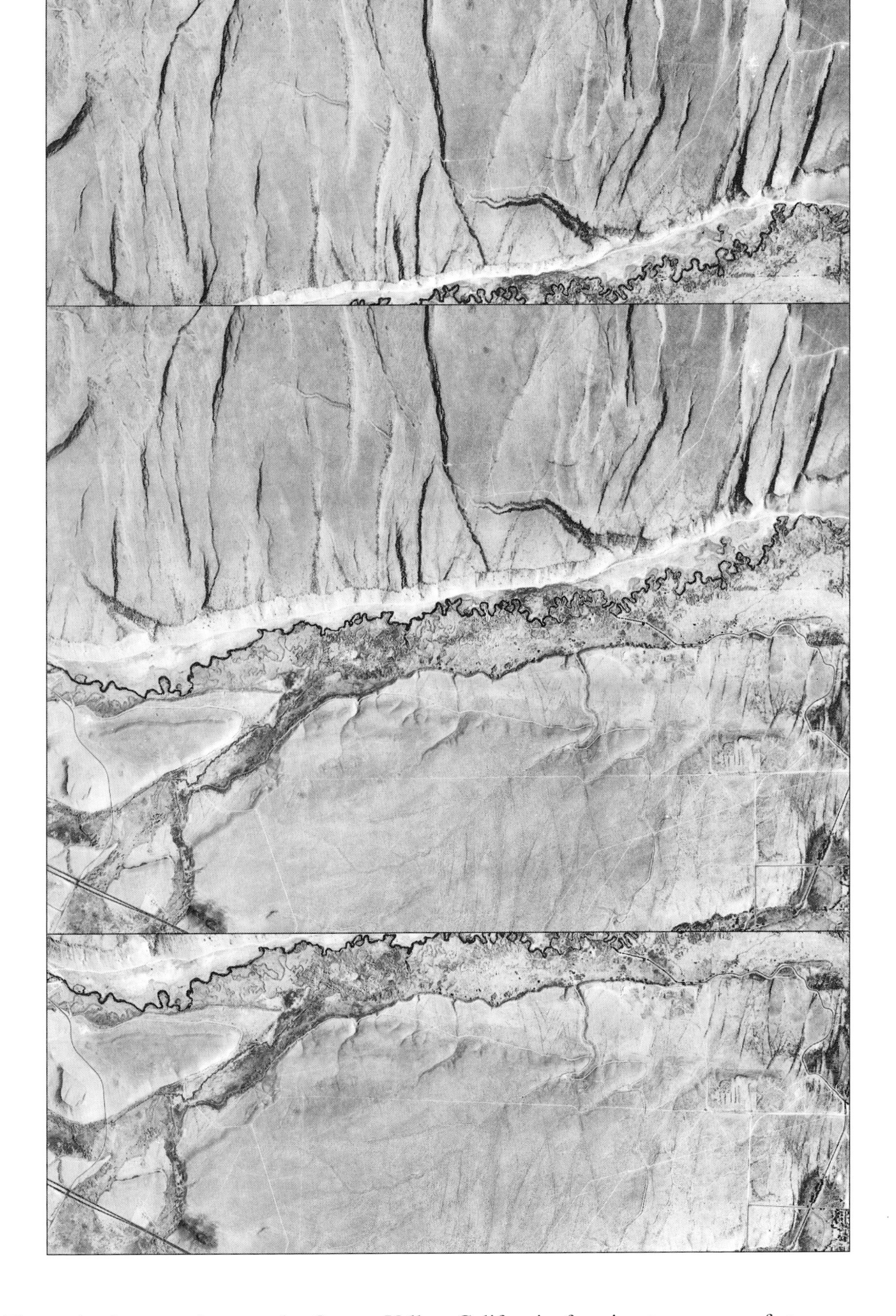

N

1:40,000

Figure 3.10 Three air photographs over the Owens Valley, California, forming two zones of stereo overlap. Scale: 1:40,000.

BIBLIOGRAPHY

Borcherdt, R.D., 1975. Studies for seismic zonation of the San Francisco Bay Region: Basis for reduction of earthquake hazards, San Francisco Bay Region, California. U.S. Geological Survey Professional Paper 941-A.

Bykerk-Kauffman, A., 1992. How faults shape the Earth. Earth, 1(6): 63-67.

Chevrier, E.D., and D.F.W. Aitkens, 1970. Topographic Map and Air Photo Interpretation. Macmillan Co. of Canada: Toronto.

Dickinson, G.C., 1979. Maps and Air Photographs: Images of the Earth. John Wiley & Sons: New York.

Easterbrook, D.J., 1993. Surface Processes and Landforms. Macmillan Publishing Co.: New York.

Pinter, N., 1995. Faulting on the Volcanic Tableland, California. Journal of Geology, 103: 73-83.

Pinter, N., and E.A. Keller, 1995. Geomorphic analysis of neotectonic deformation, northern Owens Valley, California. Geologische Rundschau, 84: 200-212.

Pinter, N., and C. Sorlien, 1991. Evidence for latest Pleistocene to Holocene movement on the Santa Cruz Island fault, California. Geology, 19: 909-912.

Spencer, E.W., 1993. Geologic Maps: A Practical Guide to the Interpretation and Preparation of Geologic Maps. Macmillan: New York.

Acknowledgements: The author would like to thank Pacific Western Co. of Santa Barbara, CA for granting permission to republish the air photographs of Santa Cruz Island.

REGIONAL FOCUS A

THE SAN ANDREAS FAULT

INTRODUCTION

The San Andreas fault is the most studied fault system in the world. To most non-scientists, California is the land of earthquakes, and the San Andreas is the one fault system that most people can identify by name. In truth, California is far from the only part of the U.S. with earthquake activity, and the San Andreas is far from the only fault in California. With those qualifications said, however, we can now say that the San Andreas fault is a regionally-continous, fast-moving, plate-bounding fault system, with among the largest, most frequent, and most regular earthquake activity of the western U.S.

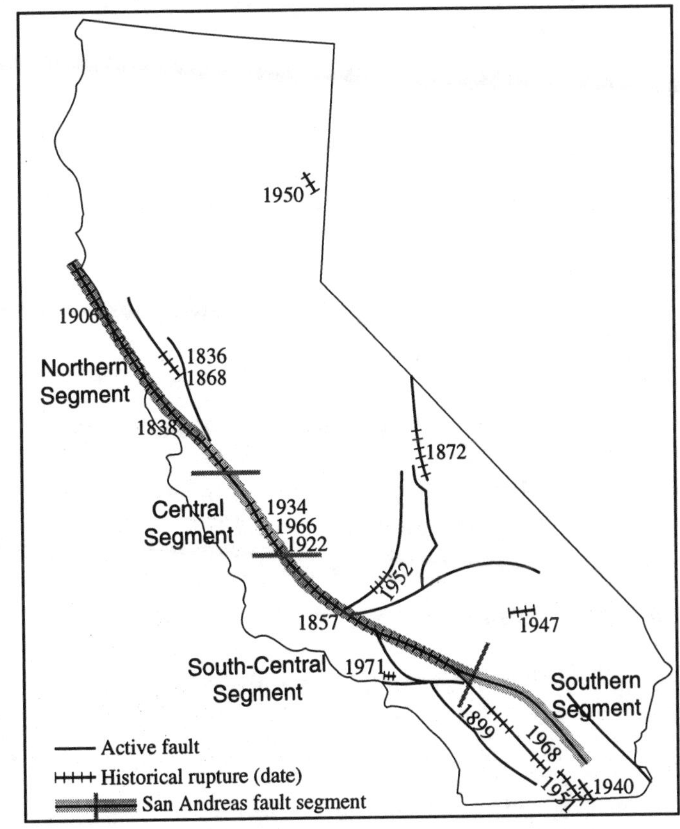

Figure A1. Location map of the San Andreas fault.

The San Andreas fault runs over 1600 km from near the California-Mexico border to the Pacific/North American/Juan de Fuca Triple Junction off the coast of northern California. The regional significance and seismic hazard associated with the San Andreas became widely recognized after the San Francisco earthquake of 1906. The 1906 earthquake (M=8.25) ruptured the fault over a distance of 400 km, shifting the land west of the fault up to 5 m northward relative to the land east of the fault (right-lateral strike-slip motion). The 1906 earthquake was the largest and most damaging earthquake on the San Andreas in historical time, but several other earthquakes have also occurred (see Figure A1).

The history of earthquake activity can be better understood by seeing that the San Andreas seems to be subdivided into discrete *fault segments.*

The San Andreas Fault

One earthquake may rupture one or more fault segments, and each segment is characterized by a different earthquake history. Both the southernmost and the northernmost segments of the San Andreas seem to be characterized by large earthquakes that occur about every 220 years (Sieh, 1984; Niemi and Hall, 1992); just northwest of Los Angeles, large earthquakes seem to occur about every 100 years (Fumal et al, 1993); the Parkfield segment has been characterized by moderate-sized earthquakes every 35 or so years; and on the Central segment of the San Andreas, constant *aseismic creep* accounts for some or all of the strain on the San Andreas.

The San Andreas fault is called the boundary between the Pacific and North American plates, but the truth is that a significant portion of the slip between the two plates occurs on faults other than the San Andreas. Distributed plate-boundary slip and many of the other geological complexities of California can be attributed to the "Big Bend" of the San Andreas. The Big Bend is the ~300 km stretch of the fault from near Palm Springs to near Santa Barbara, California that trends more east-westerly than the rest of the fault (see Figure A1). This kind of a bend on a right-lateral strike-slip fault is called a *restraining bend* because it causes localized compression. As the Pacific and North American plates have trouble slipping smoothly past each other along the Big Bend, they appear to cut subsidiary right-lateral faults to accommodate some of the motion. In addition, compression near the Big Bend has uplifted east-west trending mountain ranges in that area (the California Transverse Ranges) and is responsible for reverse faults like the one that caused the 1994 Northridge earthquake near Los Angeles. It has even been suggested that continued resistance to motion along the Big Bend is causing the current trace of the San Andreas fault to go extinct, and that a new San Andreas will form and bypass this great obstruction.

GENERAL REFERENCES

Aitken, F. and E. Hilton, 1906. A History of the Earthquake and Fire in San Francisco. The Edward Hilton Co., San Francisco.

Gore, R., 1995. Living with California's faults. National Geographic, 187 (Apr.): 2-35.

Kerr, R.A., 1993. Parkfield quakes skip a beat. Science, 259: 1120-1122.

McPhee, J.A., 1993. Assembling California. Farrar, Straus, & Giroux: New York

Mileti, D.S., C. Fitzpatrick, and B.C. Farhar, 1992. Fostering public preparedness for natural hazards: Lessons from the Parkfield earthquake prediction. Environment, 34(3): 16-20+.

Norris, R.M., and R.W. Webb, 1976, Geology of California. John Wiley & Sons: NY.

Schwartz, D.P.; and K.J. Coppersmith, 1984. Fault behavior and characteristic earthquakes: Examples from the Wasatch and San Andreas fault zones. Journal of Geophysical Research, 89: 5681-5698.

Wallace, R.E., 1990. The San Andreas fault system. U.S. Geological Survey Professional Paper 1515.

TECHNICAL REFERENCES

Crowell, J.C. (ed.), 1975. San Andreas Fault in Southern California. California Division of Mines and Geology Special Report 118.

Fumal, T.E., S.K. Pezzopane, and R.J. Weldon, 1993. A 100-year average recurrence interval for the San Andreas fault at Wrightwood, California. Science, 259: 199-203.

Michael, A.J., and J. Langbein, 1993. Earthquake Prediction Lessons from Parkfield Experiment. Eos, Transactions, American Geophysical Union, 74: 553-554.

Nadeau, R.M., W. Foxall, and T.V. McEvilly, 1995. Clustering and periodic recurrence of microearthquakes on the San Andreas fault at Parkfield, California. Science, 267: 503-507.

Niemi, T.M., and N.T. Hall, 1992. Late Holocene slip rate and recurrence of great earthquakes on the San Andreas fault in northern California. Geology, 20: 195-198.

Sieh, K., M. Stuiver, and D. Brillinger, 1989. A more precise chronology of earthquakes produced by the San Andreas fault in southern California. Journal of Geophysical Research, 94: 603-623.

Working Group on California Earthquake Probabilities, 1995. Seismic hazards in Southern California: Probable earthquake, 1994 to 2024. Bulletin of the Seismological Society of America, 85: 379-439.

DISCUSSION QUESTIONS

After reading some of the references listed above, you should be prepared to answer the following questions about the San Andreas Fault and earthquakes in California:

1) Looking at the history of large, damaging earthquakes in California, what portion of California's seismic hazard has come from the San Andreas fault? If you were in charge of allocating California earthquake-preparedness funds for the next century, what priorities would you assign?

2) How does the concept of *fault segmentation* help to explain the complex earthquake history along the San Andreas fault?

3) Explain the history of the Parkfield earthquake experiment. In retrospect, was this prediction a success, a failure, or something else?

4) Explain what role the "Big Bend" of the San Andreas fault may play in causing the very broad zone of right-lateral strike-slip faulting across Southern California and the adjacent offshore zone.

5) Explain what role the "Big Bend" of the San Andreas fault may play in causing reverse faulting in the California Transverse Ranges.

EXERCISE 4

NUMERICAL DATING

Supplies Needed

- calculator

PURPOSE

Numerical dating assigns numbers to the events and intervals of the Earth's history. Dating is often the most crucial tool in studies of active tectonics. Assessment of earthquake hazard in the future is based on information about the timing and rates of activity in the past. The purpose of this exercise is to familiarize you with the basic principles of numerical dating. You will apply these principles and use numerical dates throughout the rest of this book.

INTRODUCTION

A variety of techniques are now available to geologists studying active tectonics. In any given research situation, the technique used depends mainly on the type of material present and its suspected age. Figure 4.1 illustrates the effective age ranges for three of the most reliable dating techniques for material formed in the Quaternary (the last 1.65 million years).

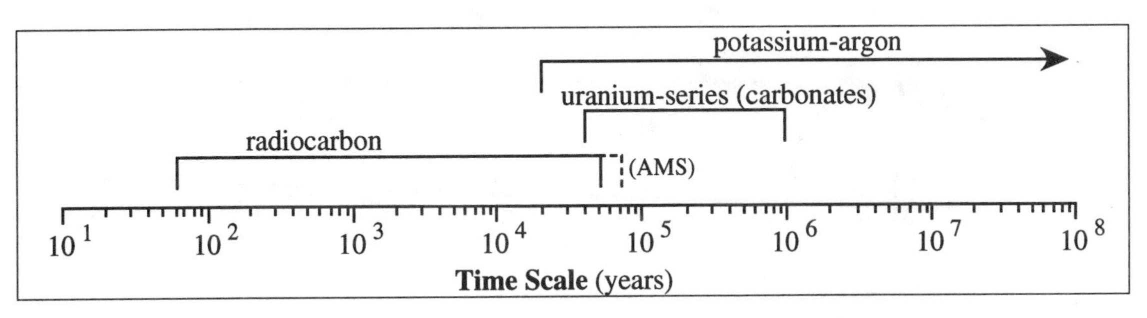

Figure 4.1. The three most-used techniques for dating material formed during the Quaternary (the last 1.65 million years) and the age ranges over which they are useful. See the text for descriptions of the types of material that can be analyzed using these methods.

Numerical Dating

The radiocarbon technique (also known as ^{14}C dating) is suitable for most organic material, including charcoal, wood, plant fiber, bone, and shell. AMS (Accelerator Mass Spectrometry) is an alternative radiocarbon technique that is more precise and requires smaller samples than does conventional radiocarbon analysis. Uranium-series dating includes several different techniques based on the decay from either ^{235}U or ^{238}U. The uranium-series technique that is most useful in geomorphology is used to determine the ages of corals and shells. Potassium-argon dating, and the more refined argon-argon technique, are suitable for dating igneous rocks and volcanic ashes. Other dating techniques useful for Quaternary material include amino-acid racemization, fission-track dating, obsidian hydration, thermoluminescence, tephrachronology, and a variety of new techniques based on cosmogenic isotopes (created by cosmic rays).

RADIOACTIVE DECAY

Most of the dating techniques listed here, including all three of the most-used methods shown in Figure 4.1, are based on measurements of radioactive *isotopes*. Each individual element in the Periodic Table has a fixed number of protons, but the number of neutrons may vary. Isotopes are forms of the same element with different numbers of neutrons and therefore different atomic mass numbers (the number of protons plus neutrons). For example, the element carbon has three different isotopes – all with six protons, but one with six neutrons (^{12}C), one with seven neutrons (^{13}C), and one with eight neutrons (^{14}C).

Carbon has two isotopes that are stable (^{12}C and ^{13}C) and one isotope that is unstable (^{14}C). The ^{14}C isotope spontaneously decays from its original form (called the *parent isotope)* into another form entirely (called the *daughter product).* The ^{14}C parent decays into ^{14}N, which is its stable and non-radioactive daughter product. Another

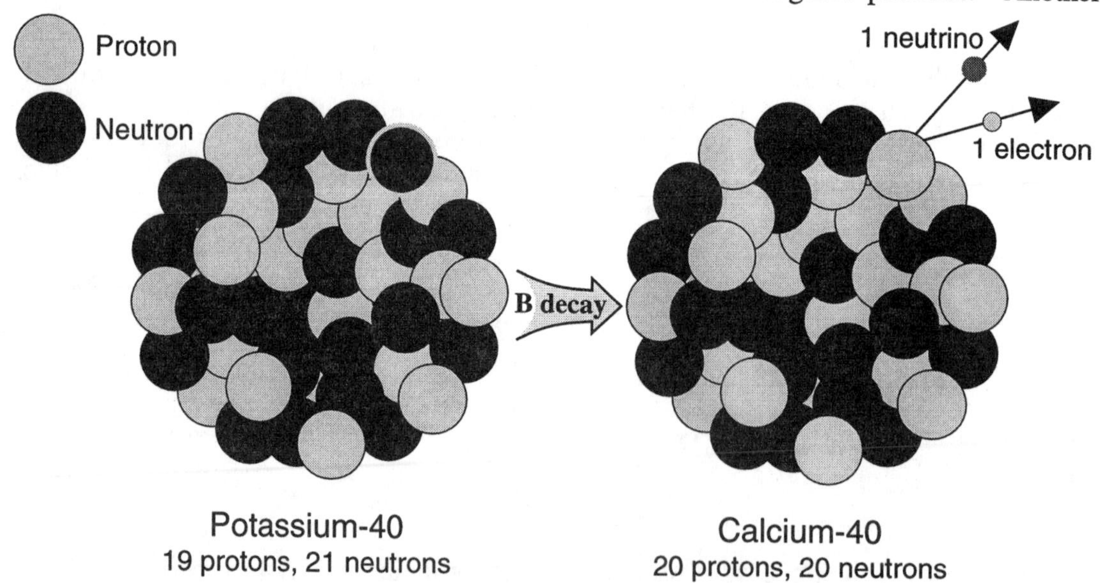

Figure 4.2. Illustration of radioactive decay. In this example, a ^{40}K is transformed into a ^{40}Ca when a neutron decays into a proton, emitting a neutrino and an electron.

example useful to numerical dating is the ^{40}K parent isotope, which has two decay paths: into ^{40}Ar and ^{40}Ca (Figure 4.2). In a closed system (for example, in a sealed mineral crystal), the number of parent atoms steadily decreases over time, while the number of daughter atoms increases (Figure 4.3). The fact that makes most numerical dating possible is that the rate of radioactive decay for a given isotope is constant. This means, for example, that by measuring the rate at which ^{40}K decays in a laboratory today, we know the decay rate throughout geological time. Decay rate of a given isotope commonly is given in terms of *half-life*, which is the time it takes for exactly one-half of the parent atoms in a closed system to turn into daughter atoms. The ratio of parents to daughters is 1:1 after one half-life, 1:3 after two half-lives (3/4 daughters), 1:7 after three half-lives (7/8 daughters), etc. (Figure 4.3).

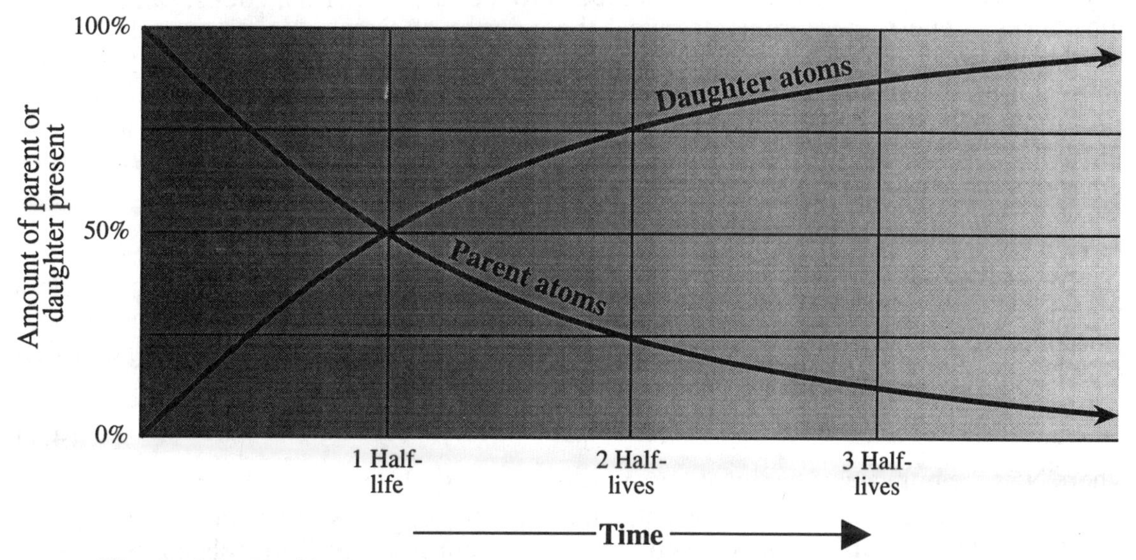

Figure 4.3. During radioactive decay, the number of parent isotopes declines, decreasing by a factor of two during each half-life. If the daughter product of the decay is stable, its abundance steadily increases.

<u>Example 4.1.</u>

As stated in the text, the element carbon consists of three different isotopes: ^{12}C, ^{13}C, and ^{14}C. In the Earth's atmosphere, the relative abundance of these three isotopes remains almost constant over time because the decay of radioactive ^{14}C is balanced by the creation of new ^{14}C by cosmic rays. All living organisms are in equilibrium with the atmosphere and have approximately the same ratio of the different carbon isotopes as the atmosphere so long as they are alive. The relative abundance of the three carbon isotopes is given below:

^{12}C	98.89%	12 amu (atomic mass units)
^{13}C	1.11%	13 amu (atomic mass units)
^{14}C	<0.01%	14 amu (atomic mass units)

Also shown above is the atomic weight of each isotope, given in atomic mass units (amu). In fact, 1 amu is defined as 1/12 the atomic weight of ^{12}C. Another way to see atomic weight is that 1 mol ($6.02 * 10^{23}$ atoms) of ^{12}C has a mass of precisely 12 grams. The

Periodic Table of the Elements lists an atomic weight of 12.011 amu for carbon, which is the average weight of the three isotopes. That value is calculated as follows:

$$\begin{aligned} 0.9889 * 12 \text{ amu} &= \quad 11.867 \text{ amu} \\ 0.0111 * 13 \text{ amu} &= +0.144 \text{ amu} \\ 0.0000 * 14 \text{ amu} &= \underline{+0.000 \text{ amu}} \\ &= \quad 12.011 \text{ amu} \end{aligned}$$

The average weight is the sum of the abundance of each isotope times its atomic weight.

Geologists make use of the systematic decay of unstable isotopes by measuring the ratio of parent to daughter atoms sealed into certain minerals, rocks, and organic substances. These measurements reveal the amount of time since the sample became sealed, knowing the radioactive decay rate. Isotopes each have their own half-lives, ranging from seconds to billions of years. Because extremely little parent material remains after more than six or seven half-lives, geologists must select an isotope with a half-life appropriate to the age of the sample. Depending on the nature of the sample and the isotope, the age estimate may represent the time when the rock or mineral formed, the time of last metamorphism (due to intense heat and pressure), or the age of the Earth.

As illustrated in Figure 4.3, the number of parent atoms decreases over time in a closed system as a result of isotopic decay. If the daughter product is stable, the number of daughters increases proportionally to the decrease in parents. The following equation describes the change in the number of parents:

$$N = N_o e^{-kt} \qquad (4.1)$$

where N is the number of parent atoms present at time t, N_o is the number of parent atoms present at t=0, and k is the rate constant. In Equation 4.1, "e" is the inverse of the natural logarithm function ("ln" on most calculators). The rate constant (k) is related to the half-life as follows:

$$t_{1/2} = \frac{0.693}{k} \qquad (4.2)$$

where $t_{1/2}$ is the half-life of the parent isotope.

Example 4.2.

Parent isotope A has a half-life of 37,000 years. If a sample that originally consisted of 100% isotope A now consists of 65% daughter product, how old is the sample? The answer to this problem is determined as follows:

The problem tells you that $t_{1/2}$ = 37,000 years. Using Equation 4.2, you can find k:

$$k = {}^{0.693}/t_{1/2} = {}^{0.693}/_{37{,}000 \text{ yrs}} = 0.000019 \text{ per year.}$$

Knowing the half-life, the original amount of parent isotope (100%) and the amount of parent isotope present now (100% – 65% = 35%), you can use Equation 4.1 to find the age of the sample:

$$35\% = 100\% * e^{-0.000019\ t}$$

$$0.35 = e^{-0.000019\ t}$$

You bring t (the sample age) down from the exponent by taking the natural logarithm of both sides of the equality:

$$\ln(0.35) = \ln(e^{-0.000019\ t})$$

$$-1.05 = -0.000019 * t$$

$$55{,}000 \text{ years} = t.$$

1) The isotopic abundance and mass of potassium are given below:

^{39}K	93.2581%	38.964 amu (atomic mass units)
^{40}K	0.0117%	39.694 amu (atomic mass units)
^{41}K	6.7302%	40.962 amu (atomic mass units)

Calculate the mass of 1.0 mol of potassium that contains the average abundance of each of the isotopes.

2) Given that the decay constant for ^{40}K is $5.305 * 10^{-10}$ per year, calculate the half-life of ^{40}K.

3) Parent isotope A decays into daughter product B with a half-life of 109,000 years. If you have an accelerator mass spectrometer that can detect as few as 1 parent isotope per 999 daughters, what is the oldest sample age that you can determine using this method? (**Hint:** Assume that the sample consisted of 100% parent isotope at t=0)

^{14}C DATING

The Earth is perpetually bombarded by cosmic rays from space. This high-energy form of radiation creates free neutrons in the upper atmosphere. The neutrons interact with nitrogen (the most abundant gas in the atmosphere), forming the ^{14}C isotope of carbon plus a hydrogen atom:

$$\text{neutron} + {}^{14}N \Rightarrow {}^{14}C + {}^{1}H. \qquad (4.3)$$

Once created in the upper atmosphere, ^{14}C combines with oxygen to form CO_2 gas, which mixes quickly throughout the whole atmosphere.

Of the three isotopesof carbon, only ^{14}C is unstable and radioactive. Over time, ^{14}C atoms decays back into the ^{14}N from which they came:

$${}^{14}C \Rightarrow {}^{14}N + \text{electron} + \text{neutrino}. \qquad (4.4)$$

This is another example of *beta decay* (see Figure 4.2). Like all unstable isotopes, ^{14}C decays at a uniform rate; its half-life is 5730 years.

All organisms incorporate carbon from the atmosphere into their bodies – plants directly from photosynthesis and vegetarians and carnivores by ingesting other plants or animals. So long as an organism is alive, it contains the same proportion of unstable ^{14}C to stable ^{12}C as the atmosphere, or as the ocean if the organism lives there. When an organism dies, however, it stops interacting with the atmosphere or ocean, and the proportion of ^{14}C in its body begins to decline as a result of radioactive decay.

When a geologist or an archeologist unearths a fragment of bone, wood, charcoal, shell, etc., they can analyze the ratio of ^{14}C to ^{12}C in the sample to determine how long ago the organism died. Two laboratory methods are used to measure this ratio. In the first method, scientists measure the number of radioactive decays in a carbon sample. For example, in 1.0 grams of *modern* carbon, 13.56 ^{14}C atoms (on average) turn into ^{14}N each minute (13.56 decays per minute, or *dcm).* In 1.0 g of 5730 year-old carbon, only one-half of the original ^{14}C remains, and only 6.78 radioactive decays would be measured. Using this technique, the age of samples as old as about 40,000 years can be determined. In the second laboratory method, scientists use an Accelerator Mass Spectrometer (AMS) to measure directly the proportion of ^{14}C atoms versus the lighter ^{12}C atoms. Using this second technique, samples as old as 70,000 years can be analyzed, and only very small quantities of carbon are required.

Radiocarbon ages are reported as a mean value (μ) and associated measurement uncertainty (σ), for example 5320±170 yBP (years before present). The uncertainty traditionally is reported as a *one standard deviation* range, meaning that there is a 68% likelihood that that actual sample age falls within that range and a 95% likelihood that it will fall within double that range. In the example of a reported age of 5320±170 years, this means that there is a 68% likelihood that the actual age is between 5150 and 5490 years BP and a 95% likelihood that the age is between 4980 and 5660 years BP. As with all radiometric results, more than one analysis of the same sample or deposit is always desirable. Multiple age estimates cannot simply be averaged, however. The method for combining multiple mean-and-standard-deviation estimates ($\mu_1 \pm \sigma_1$, $\mu_2 \pm \sigma_2$, ..., $\mu_n \pm \sigma_n$) is:

$$\sigma_{ave} = \left(\frac{1}{\sigma_1^2} + \frac{1}{\sigma_2^2} + \ldots + \frac{1}{\sigma_n^2}\right)^{-0.5} \quad (4.5)$$

$$\mu_{ave} = \sqrt{\sigma_{ave}^2 * \left(\frac{\mu_1^2}{\sigma_1^2} + \frac{\mu_2^2}{\sigma_2^2} + \ldots + \frac{\mu_n^2}{\sigma_n^2}\right)} \quad (4.6)$$

The new, properly combined age would be $\mu_{ave} \pm \sigma_{ave}$. For example, two age estimates of 970±60 (μ_1=970; σ_1=60) and 1020±100 (μ_2=1020; σ_2=100) would be combined into 983±51 years BP (μ_{ave}=983; σ_{ave}=51). Note that the combined mean age of 983 years falls between the two sample means (closer to 970 because the uncertainty of that result is small). Also, the combined standard deviation of 51 years is smaller than either of the two original uncertainties because more data usually serves to reduce uncertainty.

Example 4.3.

Example 4.2 discussed a theoretical isotope (isotope A) with a half-life of 55,000 years (k = 0.000019). You are now given the additional information that a modern 1.0 g sample of the isotope emits 4.86 decays per minute (dcm). If a 3.0 g sample of isotope A and its stable daughter product now emits 9.72 dcm, what is the age of the sample?

The key to solving this problem is that the radioactivity (dcm) of a sample is directly proportional to the amount of parent isotope present (so long as the parent is the sole source of radioactivity). If 1.0 g of modern isotope A emit 4.86 dcm, then a 1.0 g sample that is 37,000 years (one half-life) old will emit 2.43 dcm. Since radioactive emissions are directly proportional to the number of parent atoms present, then you can use that value in place of N and N_o in Equation 4.1, so long as you correct for sample weight.

If a 1.0 g sample of modern isotope A emits 4.86 dcm, then a 3.0 g sample would emit:

$$(3.0 \div 1.0) * 4.86 \text{ dcm} = 14.58 \text{ dcm}.$$

Using Equation 4.1:

$$9.72 \text{ dcm} = 14.58 \text{ dcm} * e^{-0.000019\ t}$$

Solving for t (age of the sample) using the steps outlined in Example 4.2:

$$t = 21{,}000 \text{ years}.$$

1) Calculate the rate constant, k, for the decay of ^{14}C in years.

2) You analyze a 6.5 g sample of carbon and find that it emits 2.3 beta particles per minute (dcm). What is the age of this carbon sample. (Remember that 1.0 g of modern ^{14}C emits 13.56 dcm)

Sample age = __________

3) A 5730 year-old, 1.0 g sample consists of 90% original carbon and 10% modern carbon. If you analyze the ^{14}C in this sample, what will be its apparent age? In other words, if you analyzed the sample without knowing about the contamination, what age would you get? The problem is begun for you below:

0.9 g 5730 yr carbon = _____ dcm

0.1 g modern carbon = _____ dcm

1.0 g Total Sample = _____ dcm

Actual age = 5730 yrs

Apparent age = __________

4) A 34,380 yr old sample consists of 90% original carbon and 10% modern carbon. If you analyze the ^{14}C in this sample, what will be its apparent age?

Actual age = 5730 yrs

Apparent age = ________

5) Using your results from Questions 3 and 4 above, how does sample age affect the potential for contamination in radiocarbon analysis?

6) The following age estimates are the results radiocarbon analyses, dating the age of a large earthquake on the San Andreas fault (the dates are from the Pallett Creek trench, which is the subject of Exercise 9; Sieh et al, 1989). Using Equations 4.5 and 4.6, find the combined mean and standard deviation of these radiocarbon age estimates. Also, state the age range during which you are 95% certain the earthquake *actually* occurred.

1211 ± 58 yBP
1227 ± 16 yBP
1215 ± 17 yBP
1223 ± 16 yBP

BIBLIOGRAPHY

Bard, E.B., R.G. Hamelin, and A. Zindler, 1990. Calibration of ^{14}C time scale over the past 30 ka using mass spectrometric U-Th ages from Barbados corals. Nature, 345: 405-410.

Dalrymple, G.B., and M.A. Lanphere, 1969. Potassium-argon dating: principles, techniques, and applications to geochronology. W.H. Freeman: San Francisco.

Damon, P.E., and A. Long, 1978. Temporal fluctuations of atmospheric 14C: Causal factors and implications. Annual Review of Earth and Planetary Sciences, 6: 457-494.

Easterbrook, D.J., 1993. Dating geomorphic features. In Surface Processes and Landforms. Macmillan: New York.

Pierce, K.L., 1986. Dating methods. In Active Tectonics. National Academy Press: Washington, DC.

Stuiver, M., B. Kromer, B. Becker, and C.W. Ferguson, 1986. Radiocarbon age calibration back to 13,300 years B.P. and the ^{14}C age matching of the German oak and U.S. bristle-cone pine chronologies. Radiocarbon, 28: 969-979.

EXERCISE 5

DRAINAGE BASINS AND DRAINAGE-BASIN ASYMMETRY

Supplies Needed

- calculator
- metric ruler
- felt-tip marker
- sheet of clear acetate or tracing paper

PURPOSE

Flowing water is one of the most pervasive forces on the Earth's surface, and the *drainage basin* is the fundamental unit of landscape in many studies that focus on stream systems. The study of stream systems is known as *fluvial geomorphology* (from the Latin *fluvius*, meaning "river"), but the principles of fluvial geomorphology are vital to most studies of landscape and surface processes. Earthquakes, faulting, and active deformation often influence fluvial systems, so that streams can be used to help infer the seismic and tectonic history of an area. The purpose of this exercise is to demonstrate how to map drainage basins, how to characterize different basins, and how to test for one of the simplest forms of tectonic deformation: tilt of the surface.

DRAINAGE BASINS

The network of streams that remove water from an area is called a *drainage network.* A drainage network consists of small tributary streams that feed into progressively larger and larger channels. At any given point on a drainage network, you can look upstream and delineate the area of land that drains down through that stream; this area of land is called a *drainage basin* (Figure 5.1). A drainage basin is defined as the area in which every drop of storm runoff flows through the same stream, and outside of which no drop of runoff flows through that stream. The drainage basin of a large stream contains the sub-basins of its tributary streams (see Figure 5.1). The outline of any given drainage basin, separating it from other basins, is called the *drainage divide.*

Drainage Basins and Drainage-Basin Asymmetry

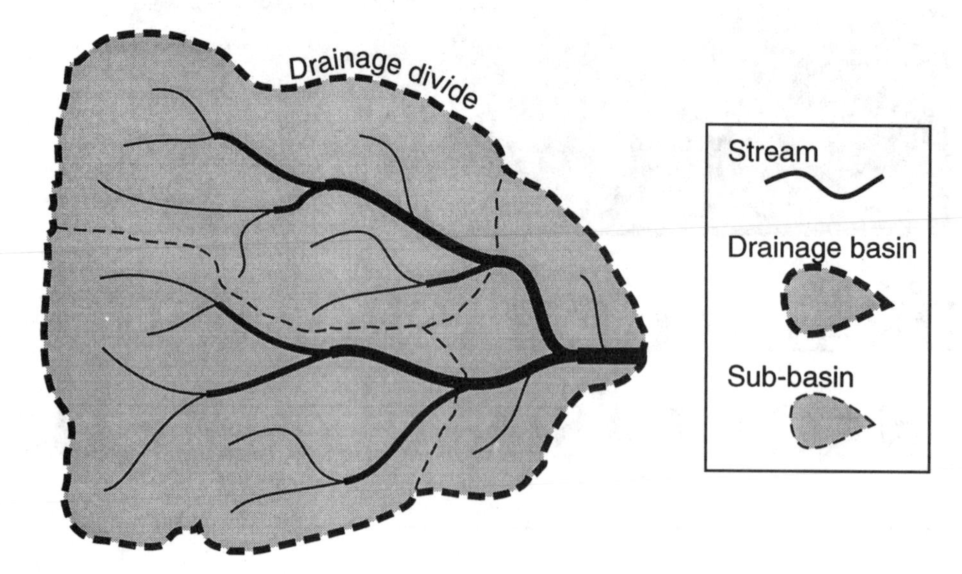

Figure 5.1. Basic characteristics of a drainage basin.

A drainage basin is defined for a single point of interest on the stream network (the point all the way on the right of the network in Figure 5.1). Any location upstream of that point would receive runoff from a smaller area. Commonly, the point of interest is the location where the stream empties into a lake or ocean, where it joins a larger river, or where it meets a mountain front. Drainage basins can be mapped on either topographic maps or on air photographs (stereo photograph pairs are very helpful, but not necessary). In order to delineate a drainage basin for a given point of interest, follow these steps:

A) Locate the point of interest on the map or photographs that you will be using.

B) Trace all tributary streams upstream from that point.

C) Extend all streams as far uphill as you can trace distinct "V"s in the topography. Add unmapped tributaries as necessary. Using stereo air photographs, extend all streams as far as you can trace a distinct channel or valley.

D) Now trace the drainage divide that encloses all of these streams. In many cases, you will be tracing a distinct ridge crest that separates this drainage basin from the adjacent ones.

1) Figure 5.2 is a stereo pair of air photographs from the west end of Santa Cruz Island, just off the coast of Santa Barbara, California. Using the four steps outlined above, delineate the drainage basin from the point marked "1". If you have an acetate sheet, mark on the acetate; otherwise, mark on the page and then make a copy on tracing paper when you're done.

Using the photo scale, what is the maximum width of this drainage basin (measured parallel to the coast)? What is the maximum length of this drainage basin (measured perpendicular to the coast)?

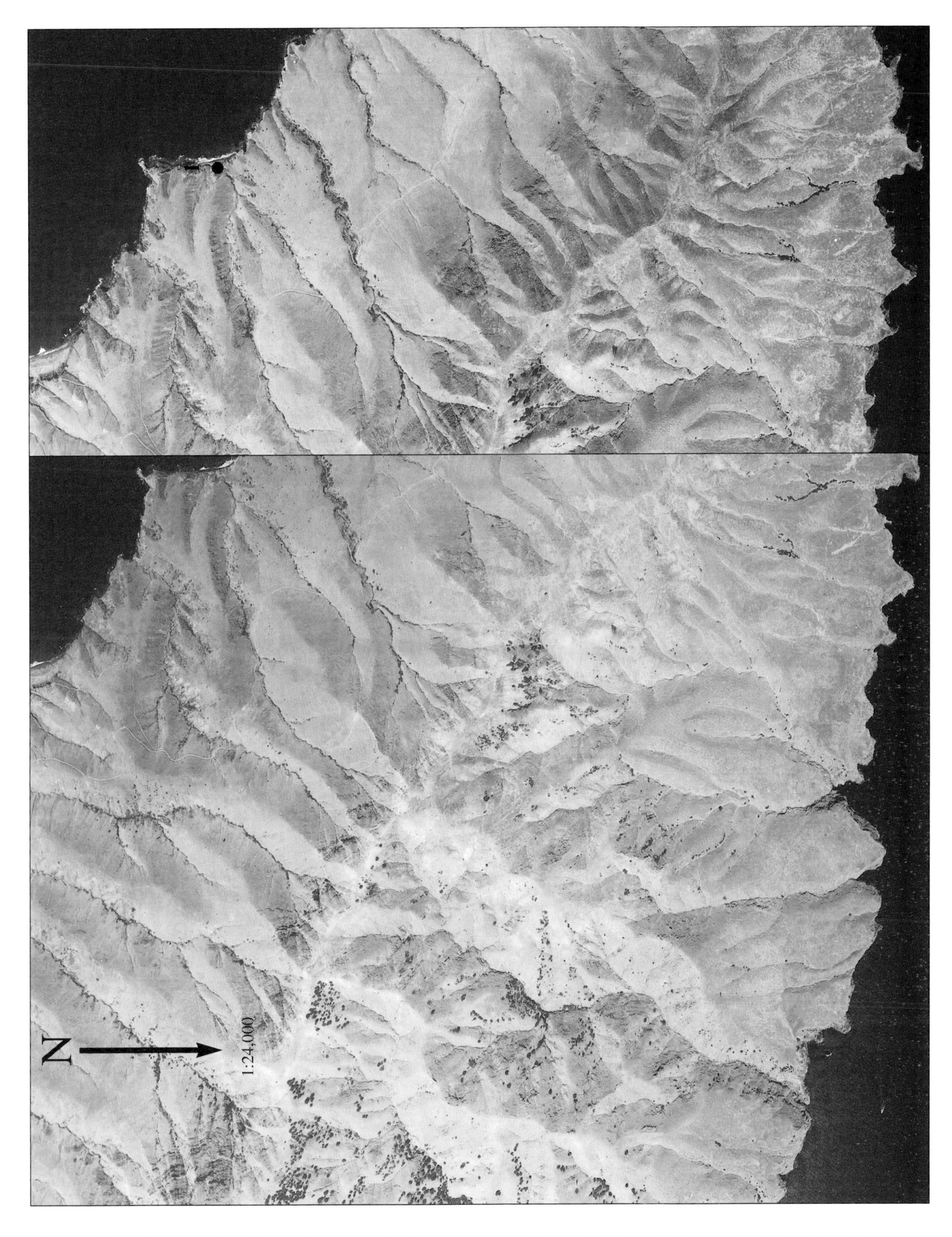

Figure 5.2 Stereo pair of air photographs from western Santa Cruz Island, California. Scale: 1:24,000. (Images courtesy of Pacific Western of Santa Barbara, CA.)

STREAM ORDER AND MAGNITUDE

A network of stream channels can be characterized in a number of ways. The importance of each stream in terms of erosional power, sediment transport, etc. is a function of how much water flows through that stream. The best way to measure the volume of water would be to measure the flow of water at an infinite number of points in the streams. Failing that, the next best way would be to measure the area of the drainage basin for each individual stream in a network (because the drainage area determines how much runoff reaches each stream). Even that compromise, however, requires many measurements. Two parameters have been developed for easily determining the relative importance of different streams in a network: the *order* of a stream (after Strahler, 1952 and 1954) and the *magnitude* of a stream (after Shreve, 1966).

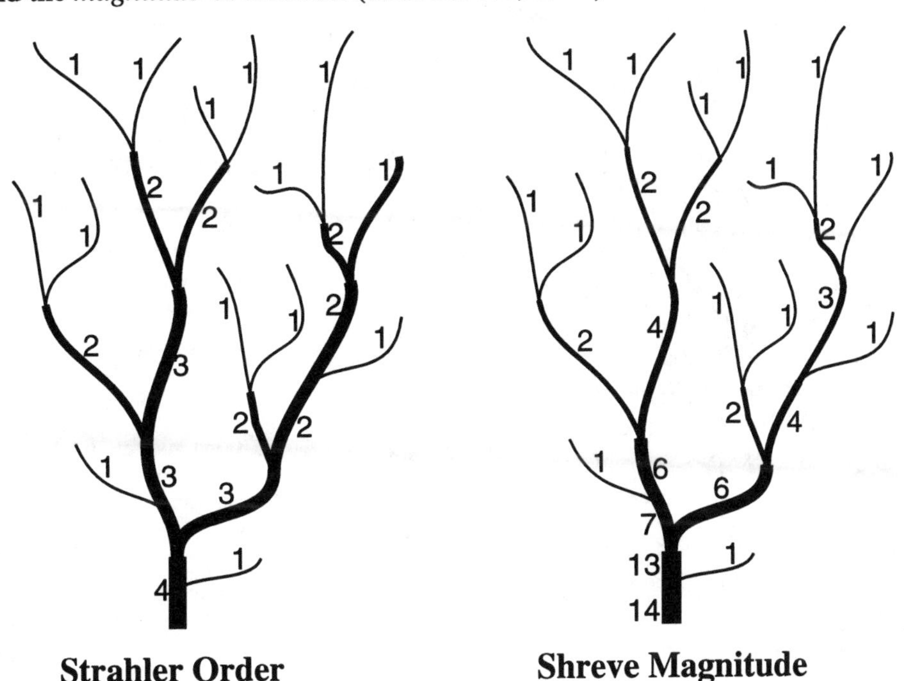

Figure 5.3. Sample drainage network for which both *order* and *magnitude* have been determined.

The smallest tributaries in a drainage network are called *first-order streams.* On the landscape, they show where runoff is first concentrated into channels. According to Strahler's system, where two first-order streams join, together they become a second-order stream. Where a first-order and a second-order stream come together, the result is still a second-order stream; but where two second-order streams join, they form a third-order stream. The rule for determining Strahler's *order* is that:

> Where two streams of the same order join, they form a stream with that order plus one. Where two different-order streams come together, the joined stream keeps the higher order of its two tributaries.

In Shreve's stream *magnitude*, the joining of a stream with magnitude 1 and a stream with magnitude 2 creates a magnitude-3 stream. Similarly, where a magnitude-3 stream and a magnitude-2 stream join, they form a magnitude-5 stream.

Drainage Basins and Drainage-Basin Asymmetry

2) Using your outline of the drainage basin on Santa Cruz Island on acetate or tracing paper, determine both the Strahler order and Shreve magnitude for each stream segment you mapped.

DRAINAGE PATTERN

The shape of drainage basins and the pattern of stream networks often given important clues about the geometry of the underlying rock and/or the nature of active tectonic deformation in that area. For example, streams can follow strong bedding planes or joints in the rocks they flow over. Stream networks can be classified into one of several possible drainage-pattern categories, some of which are illustrated in Figure 5.4.

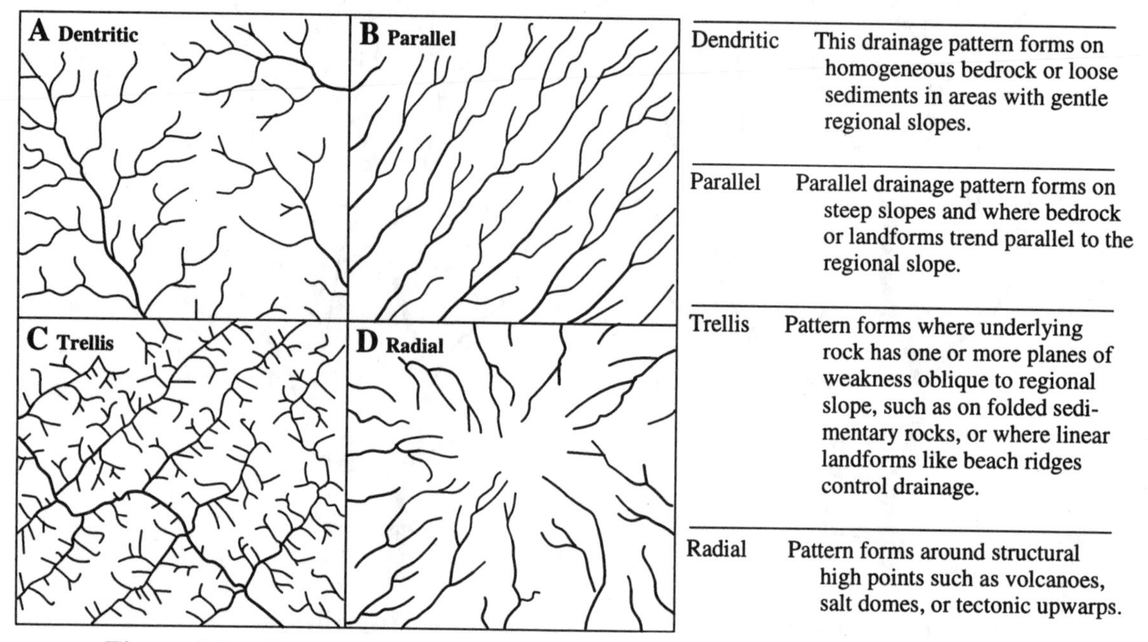

Figure 5.4. Common drainage patterns. (After Bloom, 1991)

3) What drainage pattern does the drainage basin you outlined on Santa Cruz Island have? Does this pattern indicate any strong control by rocks or rock structures?

DRAINAGE-BASIN ASYMMETRY

Where stream networks develop in the presence of active tectonic deformation, stream patterns can reflect that deformation. One of the simplest forms of deformation is tilting, which can be caused by flexure or warping of an area of the Earth's surface. Tilting can cause a stream network to become asymmetrical, with more area on one side of the drainage basin than on the other (Figure 5.5).

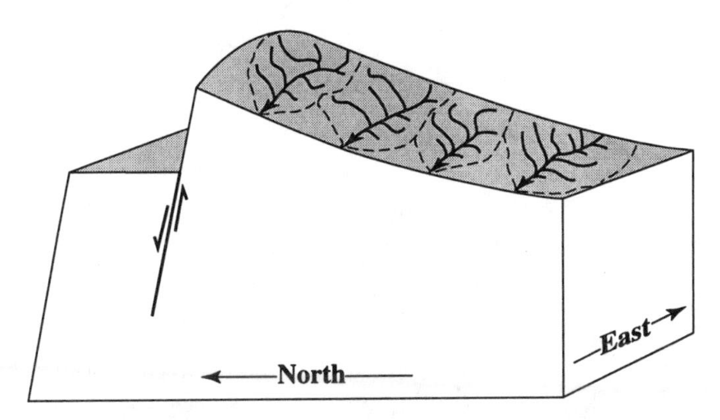

Figure 5.5. Streams flowing west across a northward upwarp show north-south asymmetry. Tilting causes tributary streams north of each main channel to elongate so that the drainage divides migrate northward. The higher the magnitude of tilt, the more pronounced the asymmetry.

Morphometric indices are techniques for measuring landforms and perturbations of landforms. Two morphometric indices are available for measuring the degree of asymmetry in a drainage basin:

A) the *Asymmetry Factor* (AF) (Hare and Gardner, 1985)

B) the *Transverse Topographic Symmetry Factor* (T) (Cox, 1994)

The **Asymmetry Factor** (AF) is defined as:

$$AF = 100 * (A_r / A_t), \qquad (5.1)$$

where A_r is the area of the basin to the right (facing downstream) of the trunk stream, and A_t is the total area of the drainage basin. For a stream network that formed and continues to flow in a stable setting, AF should equal about 50. The asymmetry factor is sensitive to tilting perpendicular to the main channel in a basin. Values of AF lower or higher than 50 may suggest tilt. For example, in a drainage basin where the trunk stream flows north and tectonic rotation is down to the west (Figure 5.5), tributaries on the east (right, facing downstream) side of the main stream are long compared to tributaries on the west side, and AF is greater than 50. If the tilting were in the opposite direction, then the larger streams would be on the left side of the main stream and AF would be less than 50.

Like most geomorphic indices, the Asymmetry Factor works best when each drainage basin is underlain by the same rock type. The method also assumes that neither lithologic controls (such as dipping sedimentary layers) nor localized climate (such as distinct vegetation on north versus south slopes) causes the asymmetry.

<u>Example 5.1</u>:

Calculate AF for the drainage basin illustrated below:

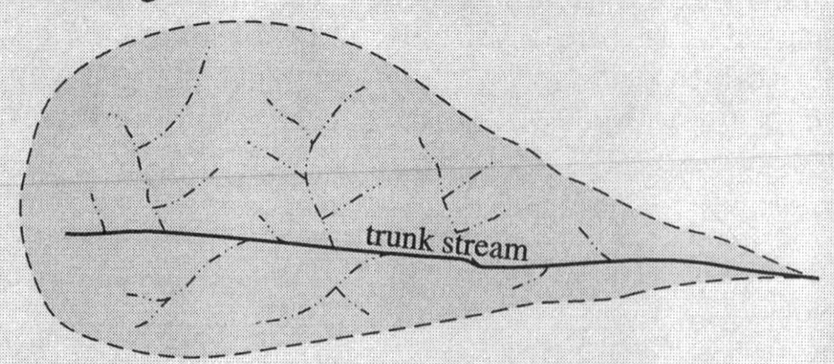

The first step here is to subdivide the drainage basin into two halves (A_r and A_t). In a basin with a single primary channel *(trunk stream)*, that channel is the dividing line. In a basin or a portion of a basin with two equal channels, you can use the ridge-line between those two channels.

The next, crucial step is to measure A_r and A_t. In this exercise, you will measure map areas by using gridded overlays. The grid below has a 0.1 km spacing. By counting the number of grid-line intersections in a given area and multiplying by 0.01 km² (because $[0.1\text{ km}]^2 = 0.01\text{ km}^2$), you can estimate the size of that area in km².

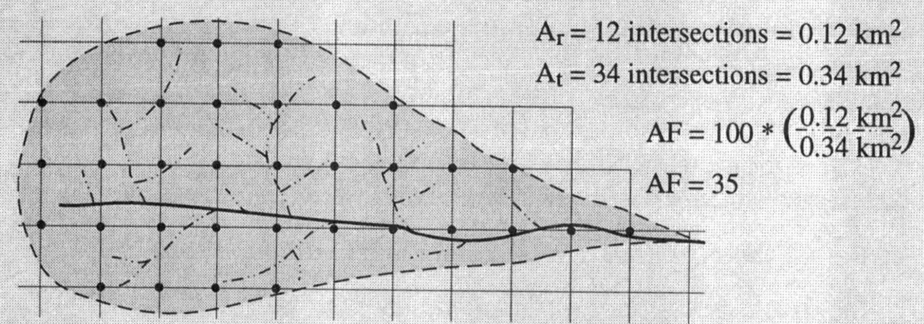

Given these measurements and using Equation 5.1, AF equals 35, indicating that the drainage basin is significantly asymmetrical.

4) What is the value of AF for the drainage basin on Santa Cruz Island? To calculate this, you will need to take your tracing of the drainage network on acetate or tracing paper and lay it over the grid in Figure 5.8 (on the last page of this exercise).

Another quantitative index to evaluate basin asymmetry is the **Transverse Topographic Symmetry Factor** (T):

$$T = D_a \div D_d \qquad (5.2)$$

where D_a is the distance from the midline of the drainage basin to midline of the active channel or meander belt, and D_d is the distance from the basin midline to the basin divide (Figure 5.6). The basin midline is defined as the line approximately parallel to the main channel that is an equal distance from the drainage divide on both sides of the basin. For a perfectly symmetric basin, T = 0. As asymmetry increases, T increases and approaches a value of 1. Values of T are calculated for different segments of valleys, and the different values can be averaged (see Figure 5.6). Like AF, T indicates preferred migration of streams perpendicular to the drainage-basin axis. This analysis is most appropriate to dendritic drainage patterns, where evaluation of tributary valleys as well as the main or trunk valley allows for a larger range of values of T.

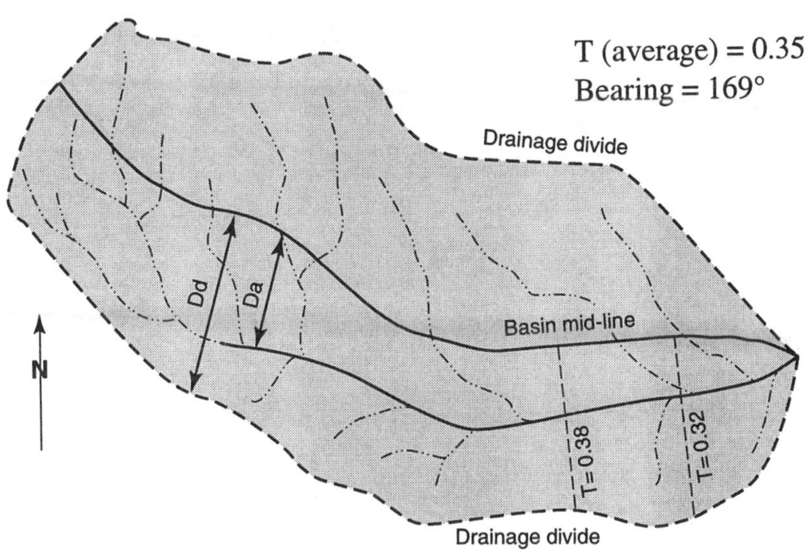

Figure 5.6. Method for calculating the Transverse Topographic Symmetry Factor (T) for evaluating drainage-basin asymmetry. (After Cox, 1994)

5) What is the value of T for the drainage basin on Santa Cruz Island? Once you have drawn the drainage-basin midline, subdivide it into 1.0 km-long segments. Calculate T for each one of these segments and average the results. Note that the trunk stream in this basin is the southern one.

Drainage Basins and Drainage-Basin Asymmetry

Both AF and T are vectors, meaning that the indices not only have magnitudes, but also directions. In Figure 5.6, the average value of T=0.35 is based on measurements oriented at 169° azimuth (11° east of south). If the average value of T were measured all the way through the basin, the average bearing could be measured perpendicular to the orientation of a line from the top of the basin midline to its bottom. Similarly, AF can be measured as a vector, the bearing of which is perpendicular to the average trend of the main channel in the basin.

Even if strong tectonic tilt does occur in a drainage basin, the drainage pattern may not be asymmetrical, depending on how the basin is oriented relative to the axis of tilt. Figure 5.7 illustrates two different drainage basins, one oriented parallel to the tilt axis and the other oriented perpendicular to it. Basin A shows the greatest asymmetry, while Basin B shows none. Intermediate cases, in which drainage basins are oriented at some oblique angle to the local tilt axis, are likely to show smaller amounts of asymmetry.

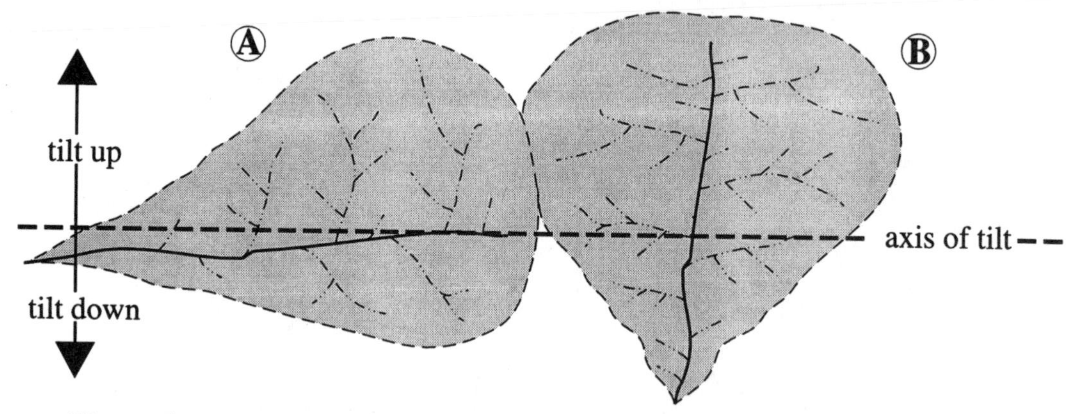

Figure 5.7. Importance of the orientation of a drainage basin relative to the local axis of deformation. Basins oriented parallel to that axis are likely to show the greatest amount of asymmetry; basins oriented perpendicular to the axis may show none.

6) Measure the average orientation of the drainage basin on Santa Cruz Island (from the highest point on the basin midline to the lowest point.

7) This drainage basin lies just north of the Santa Cruz Island fault, which is oriented at 111° azimuth (N69°W). With what type of deformation are your measurements of AF and T consistent? What orientation would you expect drainage basins with the *greatest* asymmetry to have in this area?

BIBLIOGRAPHY

Bloom, A.L., 1991. Geomorphology: A Systematic Analysis of Late Cenozoic Landforms. Prentice-Hall: Englewood Cliffs, NJ.

Cox, R T., 1994. Analysis of drainage basin symmetry as a rapid technique to identify areas of possible Quaternary tilt-block tectonics: an example from the Mississippi Embayment. Geological Society of America Bulletin, 106: 571-581.

Hare, P.W., and T.W. Gardner, 1985. Geomorphic indicators of vertical neotectonism along converging plate margins, Nicoya Peninsula, Costa Rica. In M. Morisawa and J.T. Hack (eds.), Tectonic Geomorphology. Allen & Unwin: Boston.

Shreve, R.L., 1966. Statistical law of stream numbers. Journal of Geology, 74: 17-37.

Strahler, A.N., 1952. Dynamic basis of geomorphology. Geological Society of America Bulletin, 63: 923-938.

Strahler, A.N., 1954. Quantitative geomorphology of erosional landscapes. 19th International Geologic Congress, 13(15): 341-354.

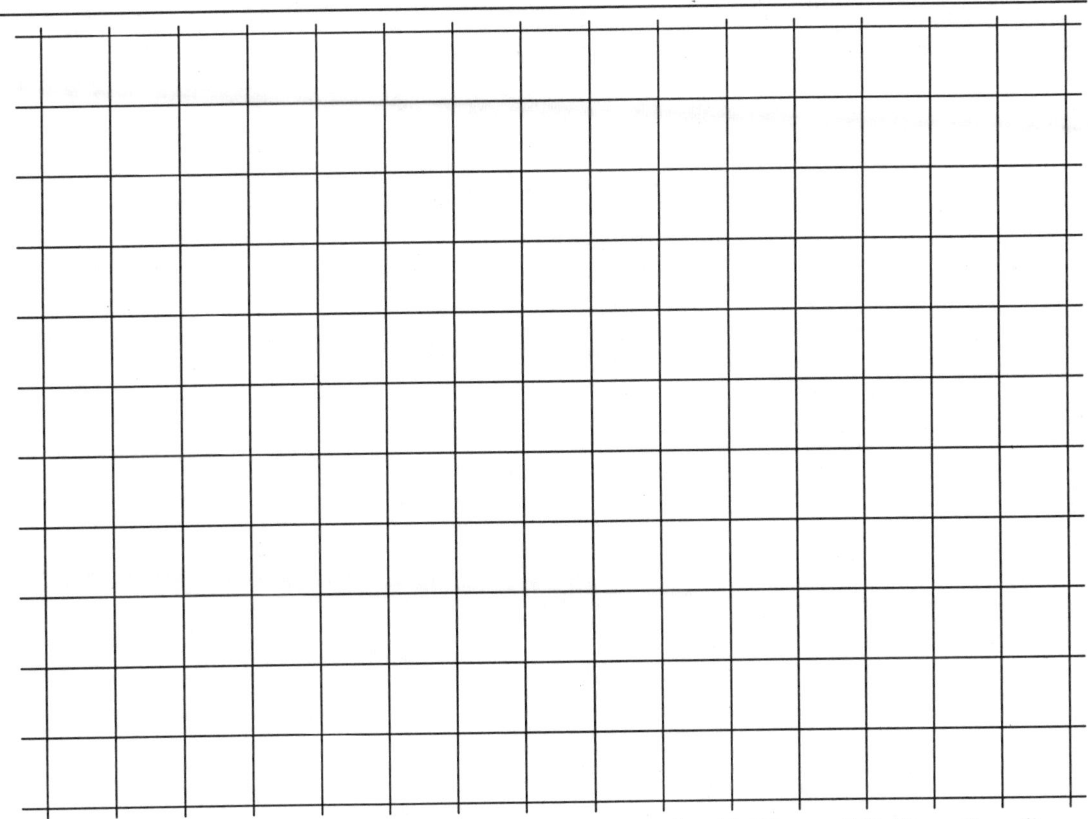

Figure 5.8. This grid is 250 m by 250 m when used with Figure 5.2 (Question 4).

REGIONAL FOCUS B

THE ATLANTIC COAST

INTRODUCTION

Many inhabitants of the eastern U.S. jealously guard one myth – they may not have the mild climate or the laid-back lifestyle of California, but at least they are spared the earthquakes. Plate tectonics adds some credibility to this myth. While the Pacific coast lies along the active margin of the North American plate, the Atlantic coast is seemingly safe in the plate's middle. Although the majority of earthquake activity does occur along the boundaries of the plates, a significant number of *intraplate earthquakes* also strike the plate interiors.

The largest historical intraplate earthquake along the Atlantic coast of the U.S. struck near Charleston, South Carolina on Aug. 31, 1886. That earthquake damaged 90% of the buildings in Charleston, destroying over 100 of them, and caused shaking as far as Boston, St. Louis, and Chicago. Intraplate earthquakes are felt, and do damage over much larger areas than plate-boundary earthquakes of the same magnitudes because seismic-wave attenuation (the loss of energy as waves travel away from their source) is less within plate interiors. It appears that near plate boundaries, the Earth's crust is more fractured, and these many fractures somehow dissipate seismic energy.

Table B1. Selected intraplate earthquakes of the U.S. and Canada.

Year	Location	Est. magnitude
1755	Cape Ann, MA	6.0
1811-12	New Madrid, MO	>8.0
1884	New York, NY	5.0
1886	Charleston, SC	7.7
1895	Charleston, MO	6.2
1897	Giles County, VA	5.8
1925	St. Lawrence River	7.0
1929	Attica, NY	5.8
1944	Massena, NY	5.6
1954	Wilkes-Barre, PA	5.0
1983	Lancaster, PA	4.3
1994	Reading, PA	5.0-5.1

Table B1 lists some of the other damaging earthquakes that have hit the eastern U.S. in the last century or so. These events, as well as the many less powerful tremors detected by seismographs, have been widely scattered, but their distribution does show several concentrations of activity (Figure B1). These *seismic zones* often are associated with much older geologic structures that seem to form zones of weakness in the crust, concentrating recent earthquake activity. One question remains, however: What driving force *causes* on-going tectonic activity in the heart of the North American Plate? A recent review of a broad range of geologic evidence (Gardner, 1989) suggests three such driving mechanisms:

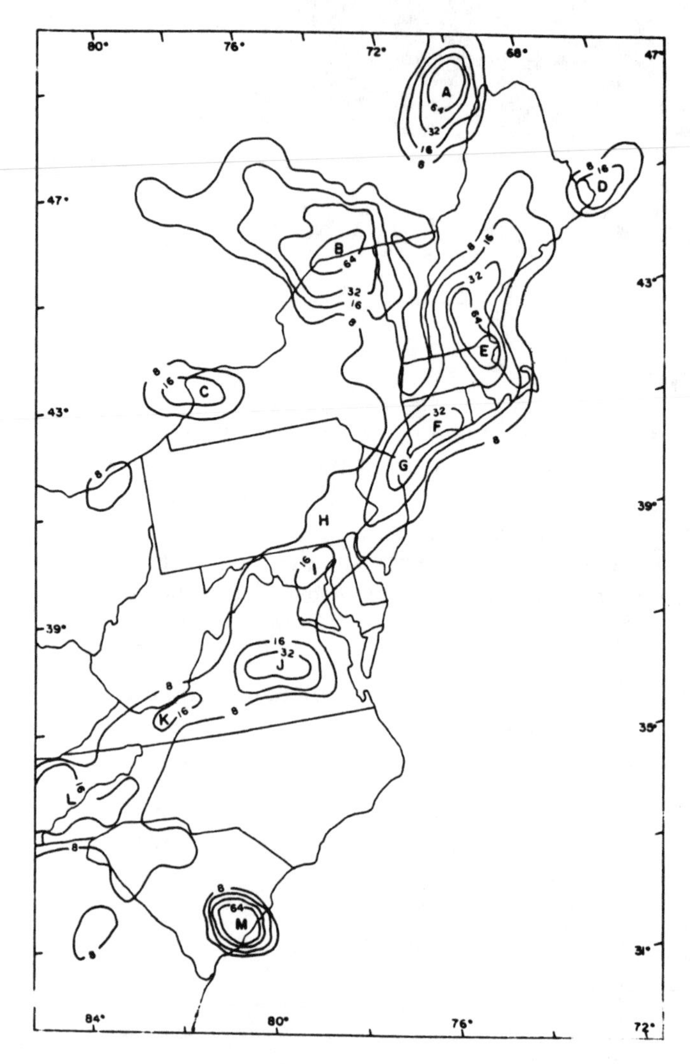

Figure B1. Map of seismic frequency in the eastern U.S. and Canada. Locations of seismic zones shown: A) Charlevoix, B) western Quebec-Ontario, C) Attica, D) Passamaquoddy Bay, E) Boston-Cape Ann, F) Moodus, G) Ramapo, H) Lancaster, I) northern Virginia-Maryland, J) central Virginia, K) Giles County, L) southern Appalachian, and M) Charleston. (From Gardner, 1989)

1) isostatic (buoyant) response to the melting of Pleistocene glaciers
2) flexure of the crust in response to erosion of the land and deposition of sediments offshore, especially in concentrated sedimentary basins
3) stress transmitted from the margins of the plate.

The interactions of these three mechanisms help to explain the complexity of intraplate seismicity and the occurrence of such earthquakes in the future.

GENERAL REFERENCES

Aggarwal, Y.P., and L.R. Sykes, 1978. Earthquakes, faults, and nuclear power plants in southern New York and northern New Jersey. Science, 200: 425-429.

Finkbeiner, A., 1990. California's revenge. Discover, 11(9): 78-82+.

Gardner, T.W., 1989. Neotectonism along the Atlantic passive continental margin: A review. Geomorphology, 2: 71-97.

Snider, F.G., 1990. Eastern U.S. earthquakes: Assessing the hazard. Geotimes, 30(11):13-15.

Stover, C.W., and J.L. Coffman, 1993. Seismicity of the United States, 1568-1989 (Revised): U.S. Geological Survey Professional Paper 1527.

TECHNICAL REFERENCES

Basham, P.W., D.J. Weichert, an M.J. Berry, 1979. Regional assessment of seismic risk in eastern Canada. Seismological Society of America Bulletin, 69: 1567-1602.

Bollinger, G.A., and R.L. Wheeler, 1983. The Giles County, Virginia, seismic zone. Science, 219: 1063-1065.

Cronin, T.M., 1981. Rates and possible causes of neotectonic vertical crustal movements of the emerged southeastern United States Atlantic Coastal Plain. Geological Society of America Bulletin, 92: 812-833.

Kuang, J., L.T. Long, an J.-C. Mareschal, 1989. Intraplate seismicity and stress in the southeastern United States. Tectonophysics, 170: 29-42.

Markewich, H.W., 1985. Geomorphic evidence for Pliocene-Pleistocene uplift in the area of the Cape Fear Arch, North Carolina. In M. Morisawa and J.T. Hack (eds.) Tectonic Geomorphology. Allen and Unwin: London.

Martin, J.R., and G.W. Clough, 1994. Seismic parameters from liquefaction evidence. Journal of Geotechnical Engineering, 120: 1345-1361.

Pazzaglia, F.J., and T.W. Gardner, 1994. Late Cenozoic flexural deformation of the middle U.S. Atlantic passive margin. Journal of Geophysical Research, 99: 12,143-12,157.

Powell, C.A., G.A. Bollinger, M.C. Chapman, M.S. Sibol, A.C. Johnston, and R.L. Wheeler, 1994. A seismotectonic model for the 300-kilometer-long eastern Tennessee seismic zone. Science, 264: 686-688.

Rhea, S., 1989. Evidence of uplift near Charleston, South Carolina. Geology, 17: 311-315.

Seeber, L., J.G. Armbruster, and G.A. Bollinger, 1982. Large-scale patterns of seismicity before and after the 1886 South Carolina earthquake. Geology, 10: 382-386.

Wheeler, R.L., and G.A. Bollinger, 1984. Seismicity and suspect terranes in the southeastern United States. Geology, 12: 323-326.

Yang, J.P., and Y.P. Aggarwal, 1981. Seismotectonics of northeastern United States and adjacent Canada. Journal of Geophysical Research, 86: 6113-6156,

Zoback, M.L., and M.D. Zoback, 1989. Tectonic stress field of the continental United States. In L. Pakiser and W. Mooney (eds.) Geophysical Framework of the Continental United States. Geological Society of America Memoir 172.

DISCUSSION QUESTIONS

After reading some of the references listed above, you should be prepared to answer the following questions about earthquakes along the Atlantic margin of the U.S.:

1) Locate three or four historical earthquakes along the Atlantic coast of the U.S. on Figure B1, and discuss their locations in terms of the pattern of seismic activity.

2) Using your knowledge of tectonic geomorphology, what kind of evidence would you look for in order to measure the long-term rate of deformation within an active seismic zone?

3) For each of the three driving mechanisms for intraplate earthquakes listed here, name the type of stress that mechanism would cause (i.e., compression, extension, shear, flexure, etc.) and the orientation of that stress (e.g., east-west extension).

4) Can you think of any other mechanisms that might deform the interior of the North American Plate and cause intraplate earthquakes?

5) In what way or ways does the relatively low frequency of earthquakes in the interior of the North American Plate *add* to the damage that earthquakes can do compared with more active areas near the plate boundary?

EXERCISE 6

COASTAL TERRACES, SEA LEVEL, AND ACTIVE TECTONICS

Supplies Needed

- calculator
- striaght-edge ruler
- graph paper

PURPOSE

Coastlines are one of the major geomorphic systems of the Earth. In addition, sea level is important to studies of active tectonics because it is a unique horizontal datum – a widespread plane of equal elevation. Because coastal landforms are created at or near sea level, finding old coastal features some distance above the modern coastline indicates tectonic deformation of the surface in the time since those features formed. Some of the most useful landforms for studies of active tectonics are coastal terraces, including both erosional terraces and uplifted coral reefs. These landforms are prominent along many coastlines, and they can be useful tools for measuring rates and patterns of tectonic uplift.

INTRODUCTION

Coastlines have been called battlefields in the war between ocean and land. Oceanic processes, including wave action and attack by chemical and biological action, fight to erode the coast and push the shore inland. The land resists the ocean's attack with the strength of coastal rock and with sand transported to the coast by rivers. Where a sandy beach protects the shore, most of the ocean's energy is consumed moving the sand around. If the supply of sediment exceeds the erosional energy of the ocean, the shore will advance ("prograde") oceanward. Tectonic uplift is another ally of the land, slowly lifting the coast. The third major force in the battle for coastlines is the worldwide position of sea level. Fluctuations of 125 m or more have accompanied the growth and decline of the great ice sheets of the Pleistocene.

Coastal Landforms

The coastlines of the Earth can be classified by the relative strength of two variables (Figure 6.1): 1) sediment supply (from sediment-rich to sediment-poor), and 2) the relative position of the land relative to sea level (including the combined effects of tectonics and sea level). As Figure 6.1 below shows, coastlines advance in response to tectonic uplift or falling sea level *(emergence)* or to sediment supply in excess of erosion *(progradation).* Coastlines retreat in response to tectonic subsidence or rising sea level *(submergence)* or to erosion in excess of sediment supply *(retrogradation,* or *shoreline retreat).*

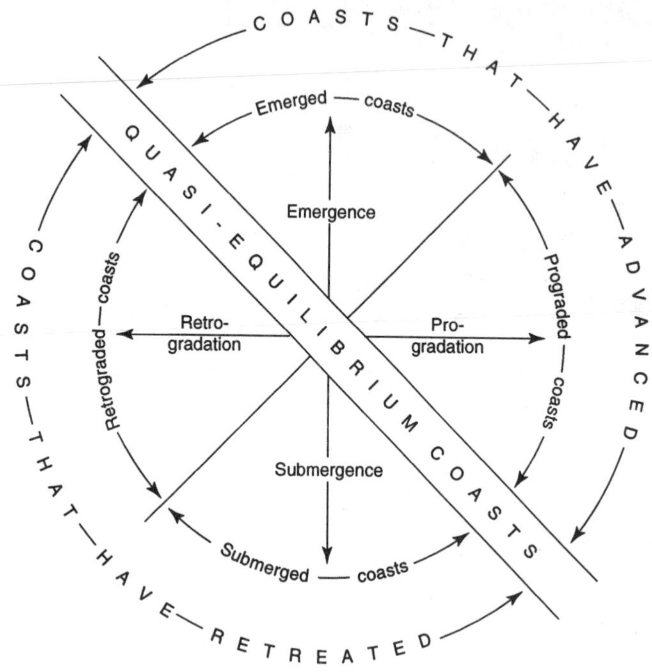

Figure 6.1. Process-oriented classification of coastlines. (After Valentin, 1970; cited in Bloom, 1991)

To isolate just one of the variables illustrated in Figure 6.1, consider coastlines that have an abundant supply of sediment. The most familiar characteristic of sediment-rich coasts is a *beach.* Tides and waves maintain a dynamic equilibrium on beaches. Storms can temporarily shift the equilibrium in favor of the ocean, stripping away large volumes of sand, but a return to low-wave-energy conditions will return sand to the beach. As mentioned earlier, sand effectively armors the shore against wave attack and erosion.

The Atlantic coast of the U.S. illustrates many of the complexities of a sediment-rich coastline. Sediment supply from rivers creates the prominent spits and barrier islands that characterize much of the coast from Fire Island, New York to Miami Beach, Florida. Barrier islands have been interpreted to be features of submergent coasts, formed as global sea level advanced in response to melting of the last Pleistocene ice sheets. Over longer periods, however, the North American coast is very slowly rising relative to the ocean in response to erosion of the land *(isostatic compensation).* Thus over glacial-to-interglacial time periods the Atlantic coast is submergent, but over longer periods it is emergent.

Landforms of sediment-rich coasts can be useful in studies of active tectonics (see for example Winker and Howard, 1977), but sediment-poor coasts generate the landforms that are most useful. The coastal landforms that are most widely used in studies of active tectonics are *coastal terraces.* Coastal terraces include two varieties: 1) erosional terraces, and 2) uplifted coral reefs. Both varieties form in close association with sea level, but erosional terraces form by wave erosion and shoreline retreat, while coral reefs form by biological activity and shoreline advance. Both erosional terraces and uplifted coral reefs reflect the complex history of glacial-to-interglacial sea-level fluctuation and long-term uplift of the land. The exercise that follows uses erosional terraces as an example, but uplifted coral reefs also are widely studied to decipher sea-level and tectonic history.

EROSIONAL COASTAL TERRACES

Erosional terraces are the characteristic landform of much of the Pacific coast of North America. In these areas, the shoreline is relatively unprotected by sediment and is exposed to marine erosion during part or all of the year. Mechanical erosion by breaking waves, as well as chemical erosion by salt and erosion by organisms that burrow into the rock, carve a broad platform just off the shore called a *wave-cut platform.* As erosion attacks the land and the wave-cut platform widens, breaking waves often cut a steep *sea cliff* into the shore. The line where the wave-cut platform meets the sea cliff is called the *shoreline angle,* and the shoreline angle is the single best marker of the position of sea level at which an erosional terrace formed. Figure 6.2 shows the features of an actively-forming erosional terrace.

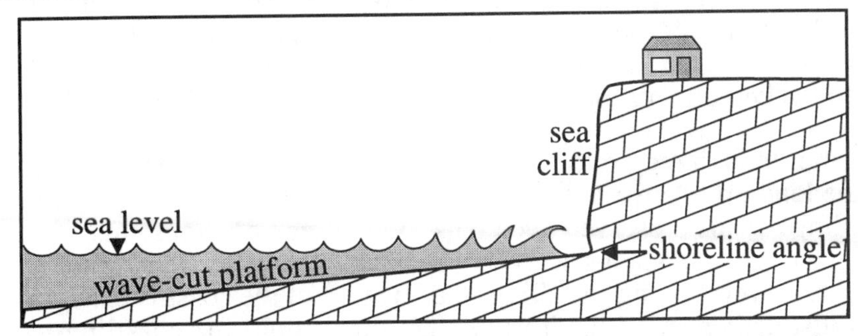

Figure 6.2. Characteristic features of an actively-forming erosional terrace.

SEA LEVEL

The ocean surface is Earth's best approximation of a perfect horizontal datum. Averaged over periods from a few hours up to a few years, sea level becomes an increasingly precise measurement of a surface of equal gravity potential. Although average sea level around the Earth has changed little in about the last 5,000 years, it fluctuated

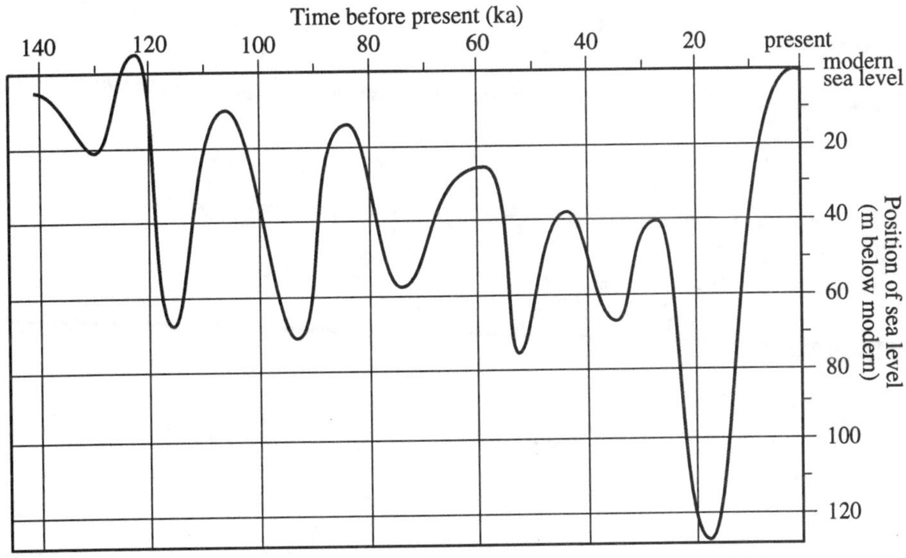

Figure 6.3. History of fluctuating sea level during the last 140,000 years.

immensely during the previous 2,000,000 years or so. Figure 6.3 is a curve that shows sea-level history during the last 140,000 years. It was inferred from uplifted coral reefs in New Guinea (Bloom et al, 1974; Pinter and Gardner, 1989). You will also use another sea-level curve (Figure 6.8) that was inferred from isotopic chemistry of fossil microorganisms. Figure 6.8 shows a longer interval of time than Figure 6.3, but the position of sea level is not precisely calibrated.

Pleistocene fluctuations in sea level primarily reflect the growth and decay of the great ice sheets that repeatedly covered much of North America and Europe. These ice sheets reached thicknesses of several thousand meters, storing up to 100 billion cubic meters of water on the continents and depleting the world's oceans proportionally. At the time of the last glacial maximum about 18,000 years ago, global sea level was approximately 125 m lower than it is today, and vast areas that are now under water were then dry land. During the last two million years, continental ice sheets grew and then melted eight to twelve times or more, and sea level has risen and fallen in synch.

Coastlines tend to record long-term fluctuations in sea level where tectonic activity slowly lifts the land up relative to the ocean. Such coasts often record multiple cycles of rising and falling oceans preserved over tens to hundreds of thousands of years. Broad wave-cut platforms are cut when sea level is near its highest position in each cycle, such as today. When sea level is rising or falling, the position of the coast is changing too rapidly to carve this distinctive set of landforms. On many uplifting coastlines, the wave-cut platform carved during the *previous* sea-level high-stand is preserved as a bench located above the modern sea cliff (Figure 6.4). These benches are known as *uplifted coastal terraces,* and staircases of as many as a dozen of such terraces are found along tectonically-

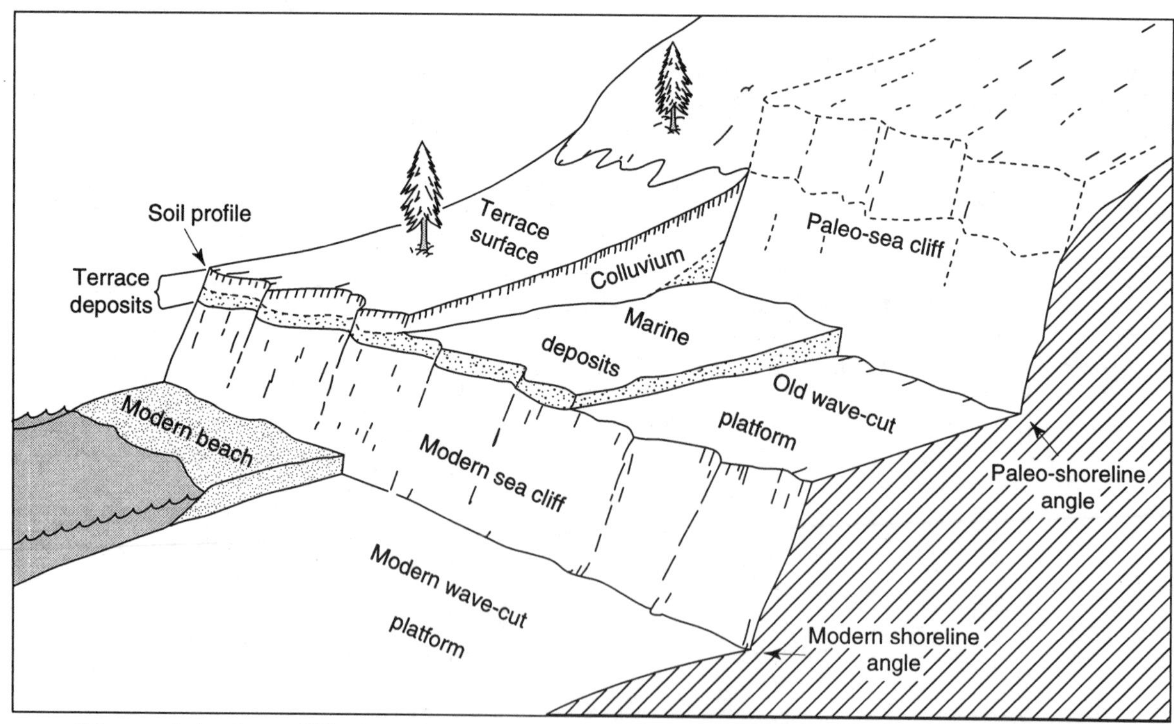

Figure 6.4. An uplifted marine terrace and associated features. (After Weber, 1983)

active coasts. You should understand that each terrace and the sea cliff above it represent the position of the shore at some time in the past. The shoreline angle measures the highest position of sea level for each interval of time. A layer of marine sediment (the last beach to cover the platform) commonly mantles the terraces, and that layer is in turn covered by a wedge of colluvium (gravity-laid debris eroded from the old sea cliff).

COASTAL TERRACES OF SAN CLEMENTE ISLAND

San Clemente Island lies 100 km off the coast of Southern California between San Diego and Los Angeles. The island is about 33 km long and a maximum of 5 km wide, forming a steep ridge more than 600 m (2000 ft) at its crest. A portion of San Clemente is illustrated in Figure 6.5 and in Figure 6.6. The island is asymmetrical, the southwest-facing slope being relatively gentle and the northeast-facing slope much steeper. This asymmetry is interpreted to reflect activity on the San Clemente fault, a right-lateral strike-slip fault which runs just off the northeast shore of the island. That fault is part of the greater San Andreas fault system, carrying a small portion of the slip between the Pacific and North American plates (see Regional Focus A).

San Clemente Island is one of the most famous locations for erosional coastal terraces in the world. At least ten different terraces cut the southwest face of the island, rising like a giant staircase from the Pacific shore (Muhs, 1983). The terraces are cut by deep river gorges and locally are covered by alluvial-fan and wind-laid deposits, but the same terrace levels can be traced along much of the length of the island. The terraces locally are covered by colluvium and by a veneer of marine sediment as illustrated in Figure 6.4 (Crittenden and Muhs, 1986). All in all, the number of terraces on San Clemente, their continuity, and their preservation are exceeded at no other site in the world.

Mapping the San Clemente Terraces and Measuring their Shoreline Angles

1) On a piece of graph paper, construct a topographic profile between points A and A' on Figure 6.5. The instructions for constructing topographic profiles were outlined in Exercise 3. Note that the topographic contours in Figure 6.5 are in feet, so that the vertical axis on the profile also will need to be in feet. In addition, you'll find that a vertical exaggeration of about five (VE=5) will nicely accentuate the terraces.

2) The lowest terrace, labeled T1, is difficult to discern on the topographic map because its elevation (5 m) is less than the contour interval of the map. Label the next terrace (the lowest one that shows on your profile) as T2, and label higher terraces sequentially from there.

3) Add the wave-cut platforms to each of the terraces on your profile. In order to do this, you'll need to know the thickness of sediment (colluvium plus marine material) on each platform. Column 2 in Table 6.1 gives this information at a number of drill-hole sites illustrated as dark dots on Profile A-A' on Figure 6.5. Assume that the wave-cut platforms are horizontal, and draw them at the appropriate elevations on your profile.

4) Add the paleo-sea cliffs to each of the terraces on your profile. The sea cliff commonly intersects the ground surface at the steepest point on the slope above an

uplifted wave-cut platform and dips steeply downward at 60° or more. On a vertically-exaggerated profile, paleo-sea cliffs can appear close to vertical. The intersection of a sea cliff and the wave-cut platform below is the shoreline angle of that terrace.

5) Measure the elevation of each shoreline angle (in ft), and enter it into Column 3 of Table 6.1. Convert the shoreline-angle elevations into meters (1 ft = 0.305 m), and enter that information in Column 4.

Table 6.1. Summary of coastal-terrace information from San Clemente Island.

1 Terrace	2 Depth to wave-cut platform (m)	3 Shoreline-angle elev. (ft)	4 Shoreline-angle elev. (m)	5 Estimated age (ka) (Questions 7, 11)
T1	0	16*	5	
T2	7			125
T3	9			
T4	11			
T5	10			
T6	15			

* see Instruction 2 above

COASTAL TERRACES, UPLIFT, AND SEA-LEVEL HISTORY

There is a *Catch 22* in studying uplifted marine terraces. You would like to know two things: 1) the ages of the terraces and 2) the long-term uplift rate in the area, but you need the age to calculate the uplift rate, and you need the uplift rate to calculate the age. The general equation for uplifted marine features is:

$$U = \frac{z - z_0 - SL}{t} \qquad (6.1)$$

where U is average uplift rate, z is modern elevation of the feature, z_0 is the elevation at which the feature formed relative to sea level at the time (depths are negative), SL is sea level at the time (see Figure 6.3), and t is the age of the feature. Equation 6.1 is a simple "rate equals distance divided by time" formula – uplift rate is the distance uplifted (present terrace elevation minus the elevation at which it formed) divided by the time interval.

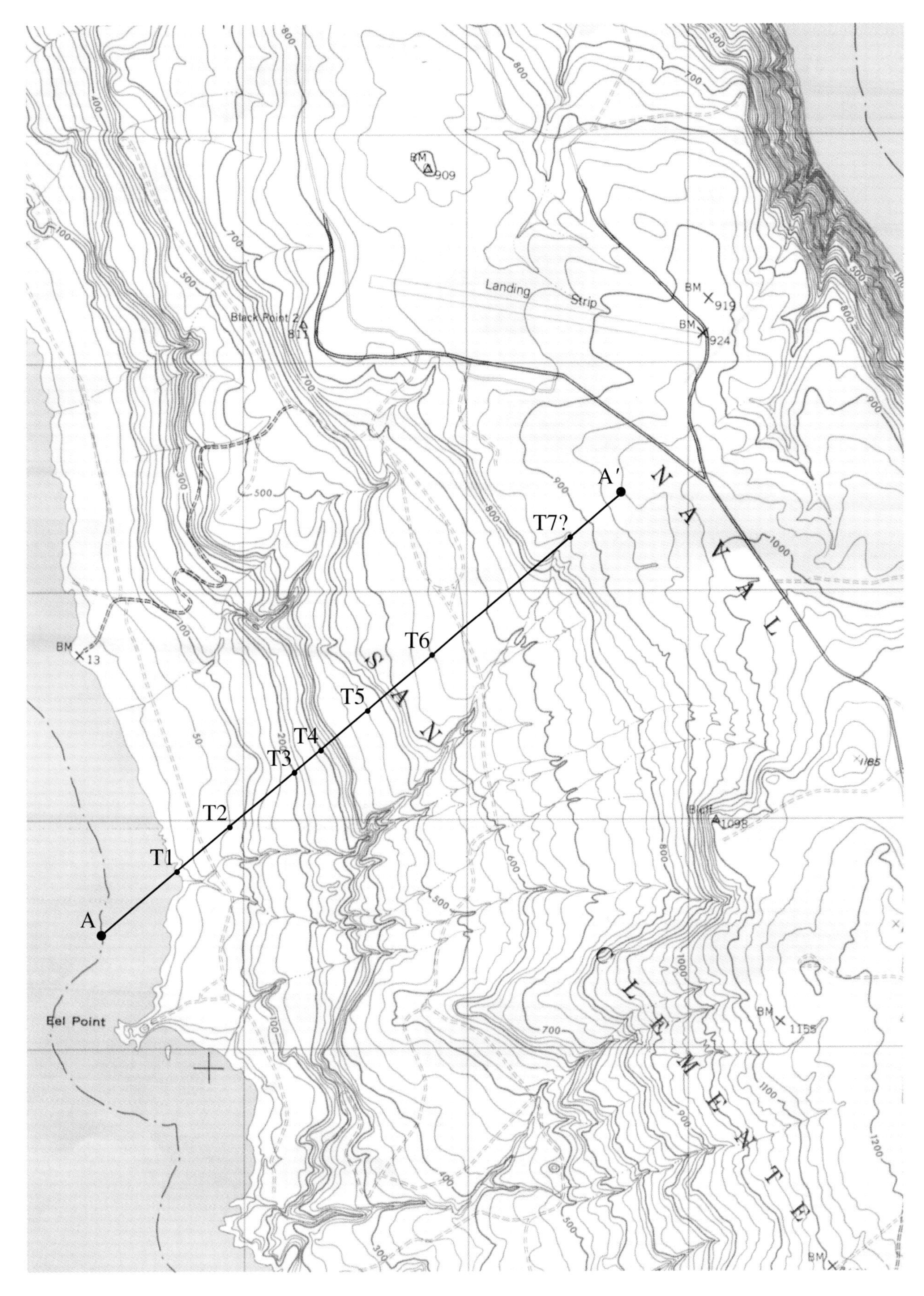

Figure 6.5 Portion of topographical map of San Clemente Island, CA.

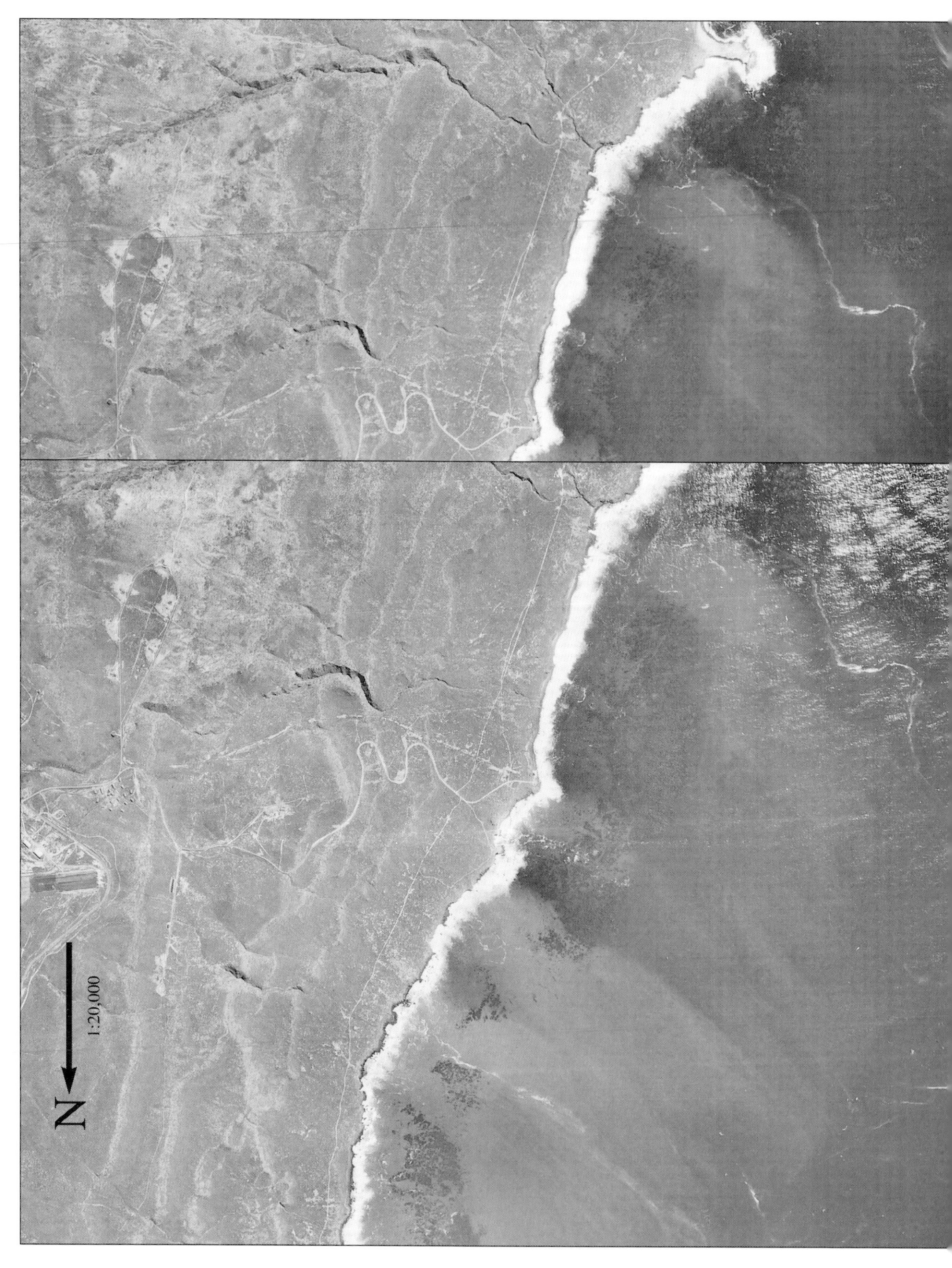

Figure 6.6 Stereo air photographs from the southwest coast of San Clemente Island, California, showing uplifted coastal terraces. Scale: 1:20,000.

Example 6.1:

Find the uplift rate at a site where you collected a 30,000 year old shell sample at an elevation of 14 m above modern sea level. The sample was found in original growth position (not transported or otherwise disturbed), and a paleontologist estimates that the organism lived about 5 m below mean sea level.

You will use Equation 6.1 to find uplift rate (U). The information above tells you that z=14 m, z_0=-5 m, and t=30,000 years=30 ka. In order to find SL, use Figure 6.3. At time t=30 ka, sea level was about 40 m below modern sea level (SL=-25 m). Plugging this information into Equation 6.1:

$$U = \frac{14\ \text{m} + 5\ \text{m} + 40\ \text{m}}{30\ \text{ka}}$$

$$U = 2.0\ \text{m/ka.}$$

Working with coastal terraces, Equation 6.1 becomes even simpler because terrace shoreline angles always form at z_0=0, so that this term drops out of the equation. Furthermore, as discussed earlier, coastal terraces form at local peaks in the sea-level curve, when sea level was at a temporary maximum. As a result, there are a limited number of ages at which coastal terraces could form (the peaks in Figures 6.3 and 6.8).

Analysis of uplifted coastal terraces is facilitated by a graphical rendering of Equation 6.1 that combines sea-level history, elevations, and uplift all in a single figure (Figure 6.7). In the example illustrated in Figure 6.7, one terrace formed at 85 ka and the other at 125 ka. The two terrace shoreline angles are now found at elevations of 66 and 131 m above sea level, respectively. On a graph like Figure 6.7, uplift rate is the *slope* of the line that connects the point at which each terrace formed with its position at t=0. The slopes of both of these lines are 1.0 m/ka. Being in the same location, the uplift for both terrace must be the same, so the two lines must be parallel. You also can use Equation 6.1 to verify this result.

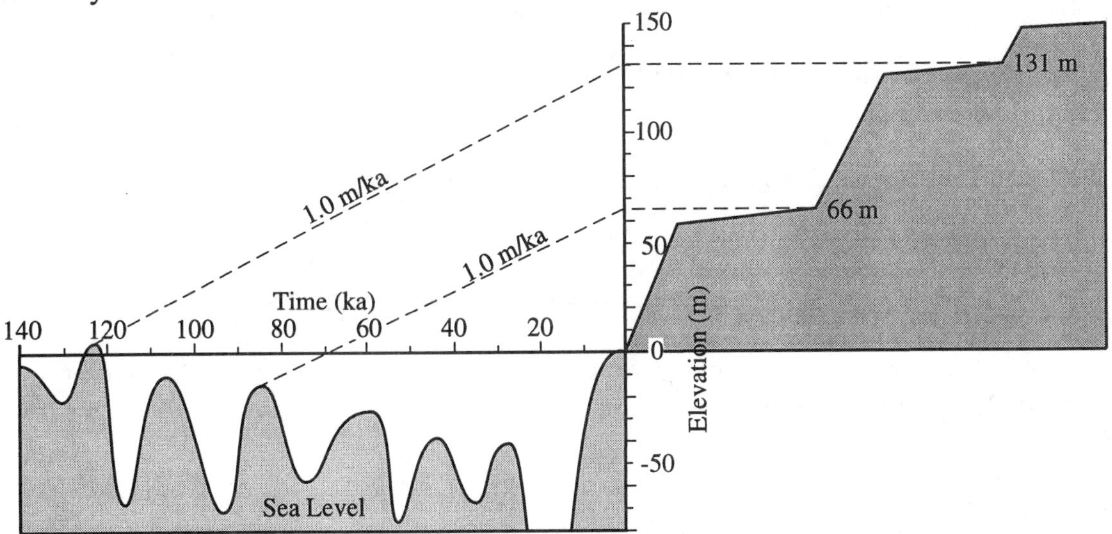

Figure 6.7. Uplift of terraces formed at 85 and 125 ka at a constant rate of 1.0 m/ka.

6) On San Clemente Island, the T2 terrace has been dated at 125 ka using uranium-series analysis of corals collected from the wave-cut platform. Using Figure 6.7 as an example, plot the position at which T2 was formed and the present elevation of its shoreline angle (at t=0) on Figure 6.8. Connect these two points. What has the average uplift rate been on San Clemente Island during the last 125,000 years?

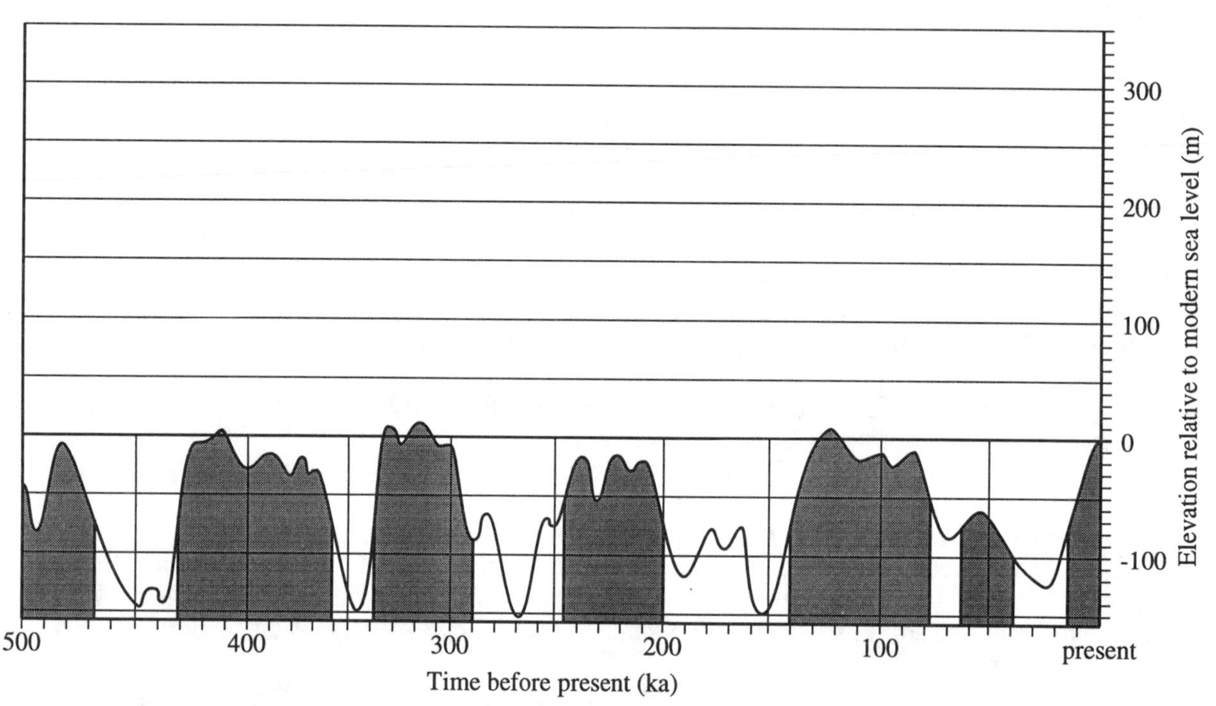

Figure 6.8. Uplift diagram for San Clemente Island terraces. (Sea-level curve after Shackleton and Opdyke, 1973)

7) Assume that the uplift rate that you calculated in the previous question has been constant during at least the last 500,000 years. Use Figure 6.8 to estimate the ages of the T1 and the T3 terraces on San Clemente Island. Follow the example shown in Figure 6.7. Enter your results in Column 5 of Table 6.1.

8) Why wouldn't Figure 6.8 be useful for estimating the ages of the highest terraces in the San Clemente sequence?

RELATIVE SPACING OF COASTAL TERRACES

You may have noticed that the assumption casually slipped into Question 7 – that an uplift rate measured in the last 125 ka has been constant during 500 ka or more – is a very sweeping claim. While it is true that uniform uplift has been demonstrated in many areas, uplift rates in other locations have varied substantially during the Quaternary (Bishop, 1991; Rockwell et al, 1988). In addition, the method outlined for Questions 6 and 7 requires that at least one terrace in a sequence can be dated using some radiometric or other numerical technique. The *relative-spacing method* is an alternative technique for determining terrace ages in sequences in which several different terraces are preserved, but numerical age control may not be present. In addition, the method tests whether the assumption of uniform uplift over time is valid or not.

The relative-spacing method requires that several terraces be present is a sequence and that shoreline-angle elevations be known with fair accuracy. This technique takes advantage of the fact that terraces seem to form only at peaks in the sea-level curve, so there are a finite number of ages to which any given terrace can correspond. If and only if there has been a constant rate of uplift, then terrace elevations can be used to infer terrace ages. The relative-spacing technique involves constructing a plot of total uplift versus time for all terraces in a sequence. Example 6.2 illustrates how the technique is used to test a hypothesis of terrace ages.

Example 6.2:

A sequence of four terraces have the following shoreline-angle elevations. Use the relative-spacing technique to test the hypothesis that the four terraces formed at the times shown:

	T1: 25 m;	T2: 55 m;	T3: 100 ka;	T4: 150 m
Hypothesis:	**t=60 ka**	**t=85 ka**	**t=107 ka**	**t=125 ka**

First of all, note that all four of the ages in the hypothesis above correspond to sea-level maxima on Figure 6.3. The next step in this technique is to calculate total uplift for each of the terraces, given the ages in the hypothesis. Completing a simple table is very helpful here.

Terrace	z (m)	t (ka)	SL at t (m)	uplift (m)
T1	2	60	- 28	
T2	29	85	- 14	
T3	75	107	- 10	
T4	183	125	+ 6	

The first three columns above reiterate the hypothesis to be tested, and the fourth column lists the positions of sea level (Figure 6.3) at the times specified. The fifth column in the table is total uplift (*not* uplift *rate*), which is simply the difference between the modern elevation of a terrace and the elevation at which it formed:

$$\text{uplift} = z - SL \qquad (6.2)$$

You should see that Equation 6.2 is just a highly simplified version of Equation 6.1 in which z_o has been removed (because shoreline angles form exactly at sea level) and in which time (t) is not considered. Completing the table:

Terrace	z (m)	t (ka)	SL at t (m)	uplift (m)
T1	2	60	- 28	30
T2	29	85	- 14	43
T3	75	107	- 10	85
T4	183	125	+ 6	177

The next step in the relative-spacing technique is to plot uplift versus terrace age (t) for all the terraces in the sequence as shown below:

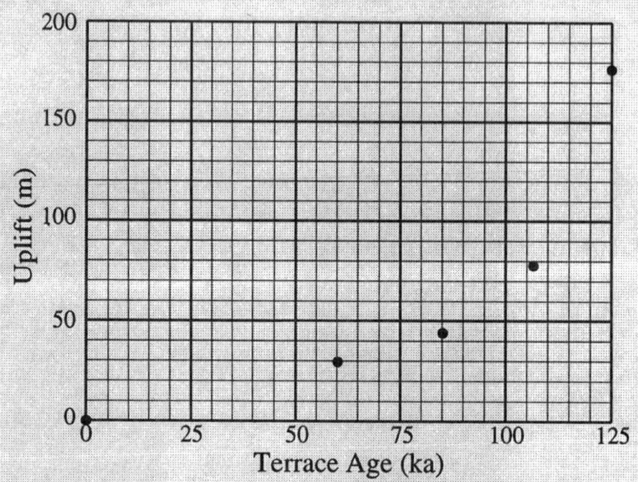

The fact that the four points on the plot above do not fall on a straight line could indicate either one or both of the following: 1) that the hypothesized ages do not work, or 2) that uplift has not been uniform over time at this site.

9) Assume that the ages for the four terraces listed in Example 6.1 are correct. Is the relative spacing of these terraces consistent with a change in uplift rate during the formation of this sequence?

What is the modern uplift rate?

What was the uplift rate before the change?

When did the change in uplift rates occur?

10) Use the relative-spacing technique for the terraces you profiled and measured on San Clemente Island to test your T1, T2, and T3 terrace dates. A data table and an uplift-vs.-time plot are provided below. Are these dates consistent with a single, constant rate of uplift during the time that these three terraces formed?

Terrace	z (m)	t (ka)	SL at t (m)	uplift (m)
T1	5			
T2		125		
T3				

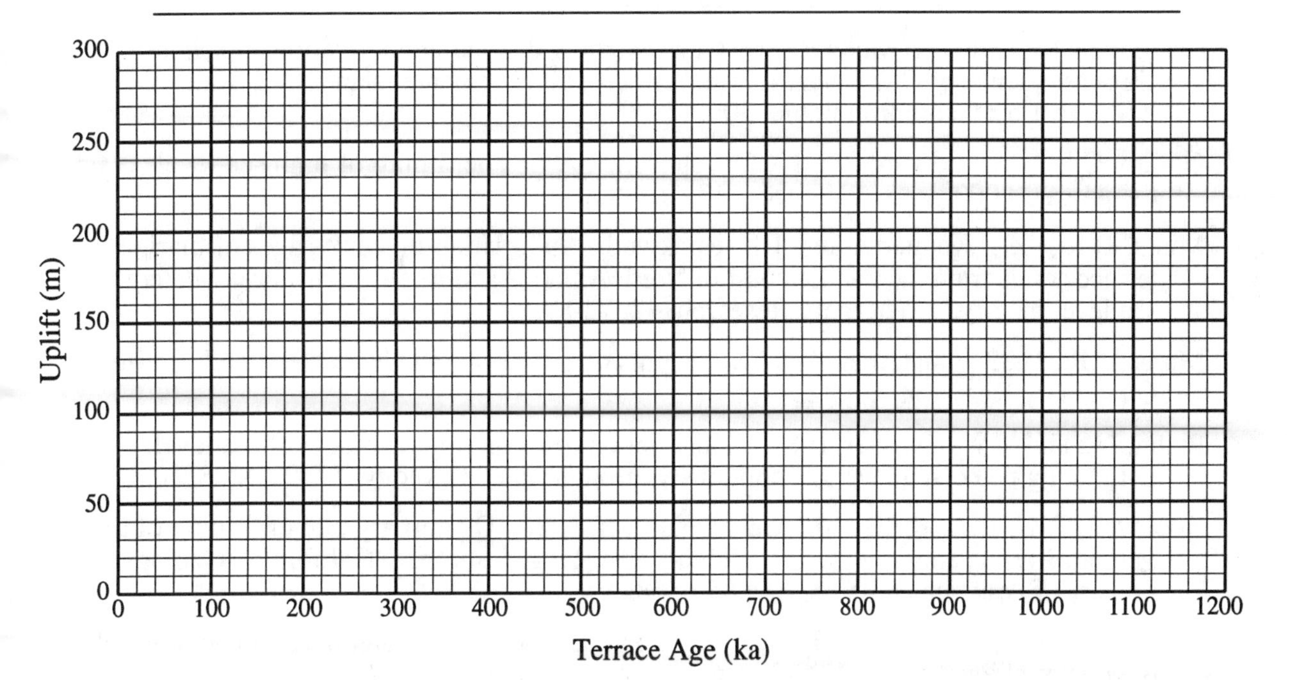

11) Now assume that all of the terraces that you profiled on San Clemente Island formed near modern sea level (SL=0; therefore uplift=z). As you can see from Figure 6.8, most interglacial high stands did fall near modern sea level. Given the present-day elevations of the higher San Clemente terraces, estimate the ages from the graph above. Enter this information in Table 6.1. What is the main assumption that these age estimates require?

Terrace	z (m)	t (ka)	SL at t (m)	uplift (m)
T4			0	
T5			0	
T6			0	

BIBLIOGRAPHY

Bishop, D.G., 1991. High-level marine terraces in western and souther New Zealand: indicators of the tectonic tempo of an active continental margin. Special Publications of the International Association of Sedimentology, 12: 69-78.

Bloom, A.L., W.S. Broecker, J.M.A. Chappell, R.K. Matthews, and K.J. Mesolella, 1974. Quaternary sea level fluctuations on a tectonic coast: New $^{230}Th/^{234}U$ dates from the Huon Peninsula, New Guinea. Quaternary Research, 4: 185-205.

Bull, W.B., 1985. Correlation of flights of global marine terraces. In M. Morisawa and J.T. Hack, (eds.), Tectonic Geomorphology. The Binghamton Symposia in Geomorphology: Internat. Series, No. 15. Allen & Unruh: Boston.

Crittenden, R., and D.R. Muhs, 1986. Cliff-height and slope-angle relationships in a chronosequence of Quaternary marine terraces, San Clemente Island, California. Zeitschrift für Geomorphologie, 30: 291-301.

Gallup, C.D., R.L. Edwards, and R.G. Johnson, 1994. The timing of high sea levels over the past 200,000 years. Science, 263: 796-800.

Muhs, D.R., G.L. Kennedy, and T.K. Rockwell, 1994. Uranium-series ages of marine terrace corals from the Pacific coast of North America and implications for last-interglacial sea level history. Quaternary Research, 42: 72-87.

Muhs, D.R., 1983. Quaternary sea-level events on northern San Clemente Island, California. Quaternary Research, 20: 322-341.

Pinter, N., and T.W. Gardner, 1989. Construction of a polynomial model of sea level: Estimating paleo-sea levels continuously through time. Geology, 17: 295-298.

Rockwell, T.K., E.A. Keller, and G.R. Dembroff, 1988. Quaternary rate of folding of the Ventura Avenue anticline, western Transverse Ranges, southern California. Geological Society of America Bulletin, 100: 850-858.

Shackleton, N.J., and N.D. Opdyke, 1973. Oxygen isotope and paleomagnetic stratigraphy of equatorial Pacific core V28-238: Oxygen isotope temperatures and ice volume on a 10^5 and 10^6 year time scale. Quaternary Research, 3: 39-55.

Valentin, H., 1970. Principles and problems of a handbook on regional coastal geomorphology of the world. Paper read at the Symposium of the IGU Commission on Coastal Geomorphology: Moscow.

Weber, G.E., 1983. Geologic investigation of the marine terraces of the San Simeon fault zone, San Luis Obispo County, California. Technical Report to the U.S. Geological Survey, Contract #14-08-001-18230.

Winker, C.D., and J.D. Howard, 1977. Correlation of tectonically deformed shorelines on the southern Atlantic coastal plain. Geology, 5: 123-127.

Acknowledgements: All San Clemente Island data is from Muhs (1983) and Muhs et al (1994), with the exception of the drill-hole information (Table 6.1, Column 1), which was fabricated for this exercise.

EXERCISE 7

BALANCED CROSS SECTIONS AND RETRODEFORMATION

Supplies Needed

- calculator
- metric ruler
- scissors
- tape

PURPOSE

Deformation of the Earth's surface is one of the most visible results of active tectonics, but it is not the whole story. Some faults ("buried reverse faults") can cause large earthquakes (for example, the 1994 Northridge earthquake that struck the Los Angeles region) even though they never break the surface. Two of the most useful tools for studying faults that do not break the surface are 1) balanced cross sections, and 2) retrodeformation. This exercise will acquaint you with these two tools and show you how they can be applied to problems in active tectonics, active folding, and earthquake hazard.

BALANCED ("Retrodeformable") CROSS SECTIONS

Geological structures include folds, faults, joints, and other evidence of deformation within the Earth's crust. The majority of such structures around the world formed in the ancient geologic past, but others are actively forming today. Cross sections through the crust in zones of active deformation often reveal that layers of rock and other geologic features mimic patterns of deformation at the surface, but show much more dramatic change because the rocks usually are much older than surface features.

Geological structures are vitally important in the field of petroleum exploration. Faults and folds are responsible for trapping oil and gas in concentrated pockets. Exploration geologists spend a lot of their time trying to locate these kinds of structures. The petroleum industry uses a variety of sophisticated techniques to probe the subsurface, including exploratory drilling that can determine rock type and bedding orientation several kilometers below the surface, and seismic-reflection profiling that determines large-scale relationships below the surface. This type of information is extremely useful in studies of active tectonics.

Balanced Cross Sections and Retrodeformation

The key to interpreting geologic structures, whether active or inactive, is the construction of *cross sections* that show the types of rock present beneath the surface; the orientations of bedding at different locations and depths; and the presence of folds, faults, and other geologic structures. Geologists typically have only localized data, and they must extrapolate across data-poor areas in order to construct regional cross sections. Such extrapolation is not haphazard, but is based on models of how geologic structures should look. A geologic cross section is said to be *balanced* if the thickness of each sedimentary layer is the same throughout the profile. To construct a balanced cross section, follow these steps:

A) Construct a topographic profile along the line of section. Typically, a cross section shows deformation most clearly if it is oriented perpendicular to the strike or trend of the structures. In addition, the profiles should have no vertical exaggeration unless bedding and structures in the area are extremely subtle. An unexaggerated profile shows true bedding thicknesses regardless of the dip of the layers. Without vertical exaggeration, the surface topography will look quite gentle on most profiles.

B) Add the geology visible at the surface, including rock types, boundaries between different units, bedding dips at different locations, and the location of any faults that rupture the surface. This information can come from geologic maps and/or field mapping.

C) Add information from wells.

D) Extrapolate between locations where you do have information to where you do not. In general, it's best to begin with stratigraphic contacts, infer folds from changes in dip, and then add faults where abrupt breaks in the sequence are indicated. The *Kink Method*, outlined in Exercise A, is one technique for systematically extrapolating dips across a cross section.

E) Check if the cross section is *retrodeformable.*

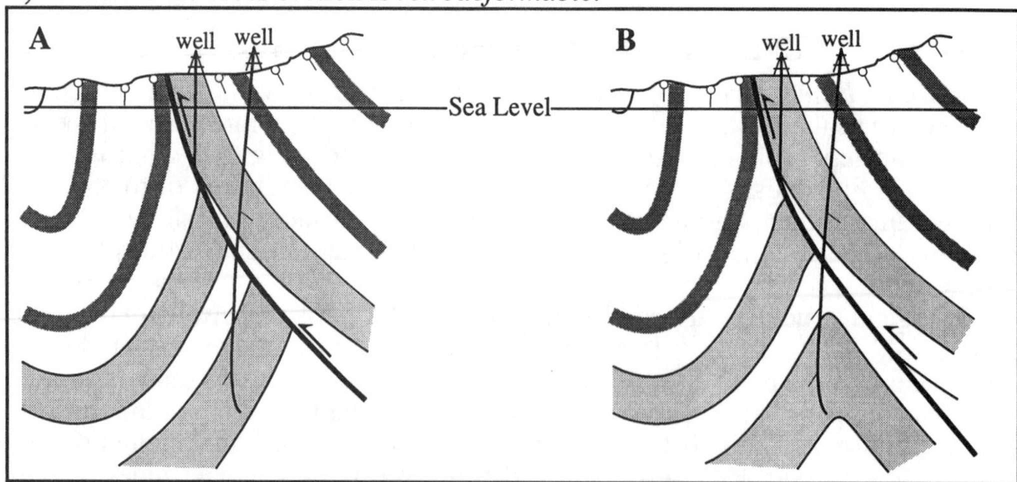

Figure 7.1. Two interpretations of surface geology and well data on a faulted anticline in western Taiwan. Interpretation A is not retro-deformable; Interpretation B is. (After Suppe, 1985)

Interpretation of cross sections can be subjective, especially where the hard data is sparse. Step E on the previous page is crucial because it checks whether the subjective processes of extrapolation and interpretation result in a model that is feasible. Presumably, all sequences of rocks originally were undeformed before they were folded and faulted. If the sequence was deformed, then at least on paper and in the human imagination, it also can be *undeformed.* Figure 7.1 shows two interpretations of the same cross section, one that is not retrodeformable and one that is. The process of checking retrodeformability is outlined later in this exercise.

THE KINK METHOD OF EXTRAPOLATION

Extrapolating bedding orientations across zones without data need not be a random act. Fortunately, even intensely deformed rock layers follow certain rules. The shape of one stratum is constrained by the shape of the strata above and below it. The Kink Method makes two assumptions about folded strata:

A) that folds are parallel, which means that the thickness of any given stratum (measured perpendicular to bedding) is locally uniform,

B) that folds consist of zones of uniform dip that are separated by distinct *kinks* (see Figure 7.2 below).

The second assumption above commonly is found to be true in the field. Even where folds consist of broad curves, the curves often consist of many small, straight-line segments. The kinks in different layers are connected to kinks in underlying layers along lines called *axial surfaces* (the lines on the cross section are surfaces in three dimensions). Each axial surface bisects each kink, meaning that it cuts the angle of the kink exactly in half (for example, you can see that $\gamma_1=\gamma_2$ and $\gamma'_1=\gamma'_2$ on Figure 7.2).

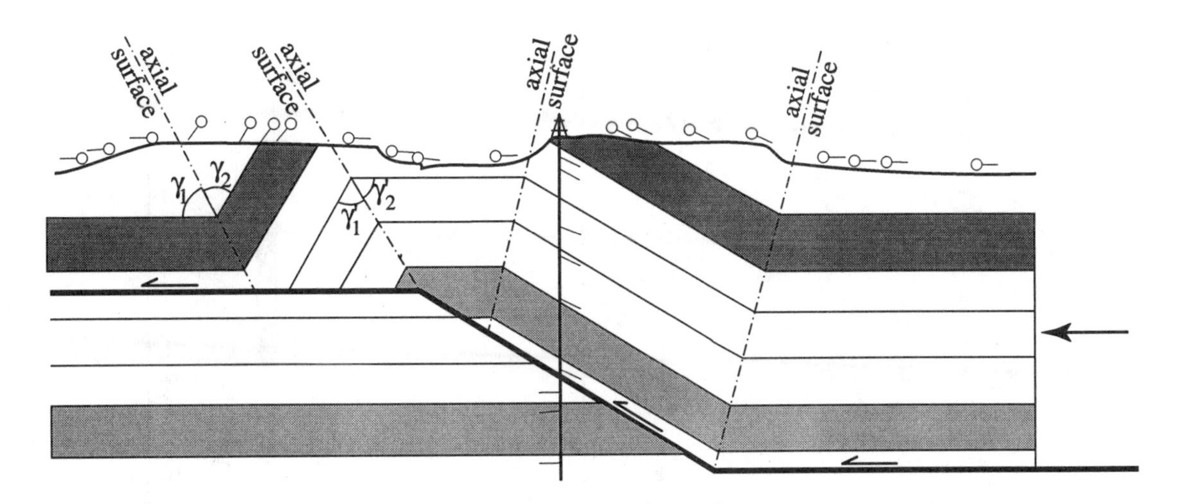

Figure 7.2. An anticline that consists of four kinks. The anticline grew as a result of slip on the underlying fault ramp. (After Suppe, 1983)

Balanced Cross Sections and Retrodeformation

1) Use the rules of balanced cross sections and the Kink Method to complete this section:

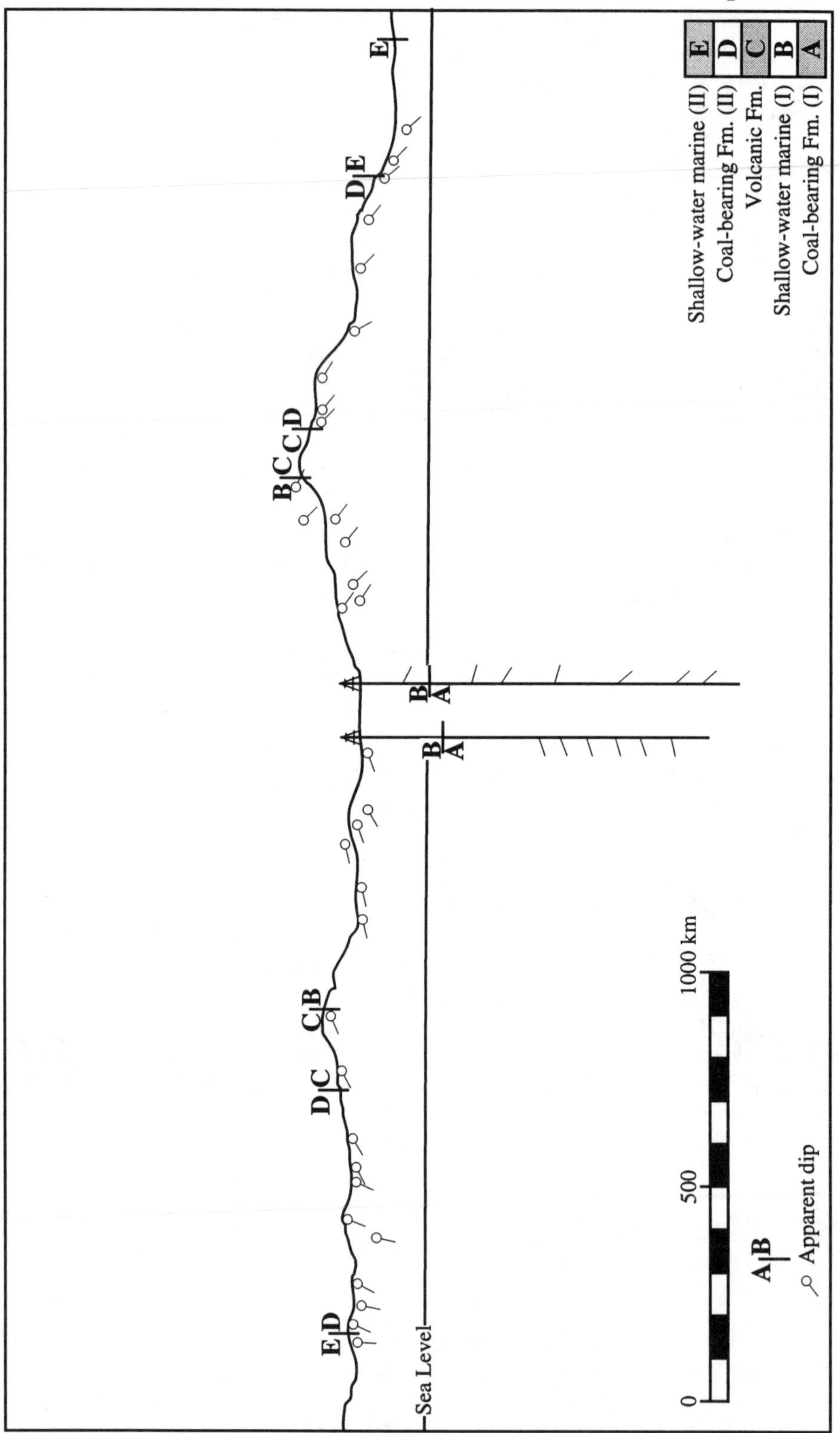

Figure 7.3. Unfinished cross section of the Shantzechiao anticline, Taiwan. (After Suppe, 1985)

RETRODEFORMATION

Figure 7.4 is a balanced cross section of a portion of the Taiwan fold-and-thrust belt. You will test whether this interpretation of the surface geology and the well-hole data is retrodeformable, and if it is, you will measure the amount of deformation that has occurred. Follow these steps:

A) Plot the axial surfaces of the folds onto the cross section.

B) Using scissors, cut the cross section into pieces by cutting along every fault plane and every axial surface. Place the pieces on an empty desk or tabletop.

C) Undeform the sequence. You'll do this by rearranging the pieces so that the different strata are in the right order and the thickness of each stratum is uniform or varies systematically. As you place each piece in its undeformed position, tape it down on your tabletop. Note that small gaps are unavoidable at the kinks; in the rocks, slip parallel to bedding or other internal deformation occurred in these locations.

D) Answer the following questions:

2) What distance has Fault A slipped? To get distance along the fault plane, remember that this cross section has no vertical exaggeration (VE=1).

3) What is the minimum distance that Fault B has slipped?

4) Why is your answer above a *minimum* distance? (i.e.: What happened to the portion of the layer right above Fault B where the words "Fault B" are printed?)

5) Calculate the minimum amount of *crustal shortening* across this profile as a result of faulting and folding. You can calculate shortening as follows:

*(width of the profile before retrodeformation) ÷ (width of the profile after retrodeformation) * 100%*

BIBLIOGRAPHY

Davis, T.L., J. Namson, and R.F. Yerkes, 1989, A cross section of the Los Angeles Area: Seismically active fold and thrust belt, the 1987 Whittier Narrows earthquake, and earthquake hazard. Journal of Geophysical Research, 94: 9644-9664.

Davis, T.L., and J.S. Namson, 1994, A balanced cross-section of the 1994 Northridge earthquake. Southern California, Nature, 372: 167-169.

Mount, V.S., J. Suppe, and S.C. Hook, 1990. A forward modeling strategy for balancing cross sections. American Association of Petroleum Geologists Bulletin, 74: 521-531.

Namson, J., and T. Davis, 1988. Structural transect of the Western Transverse Ranges, California: Implications for lithospheric kinematics and seismic risk evaluation. Geology, 16: 675-679.

Suppe, J., 1983. Geometry and kinematics of fault-bend folding. American Journal of Science, 283: 684-721.

Suppe, J., 1985. Principles of Structural Geology. Prentice-Hall: Englewood Cliffs, New Jersey.

Acknowledgements: The author would like to thank Dr. John Suppe of Princeton University for his permission to use the figures in this exercise.

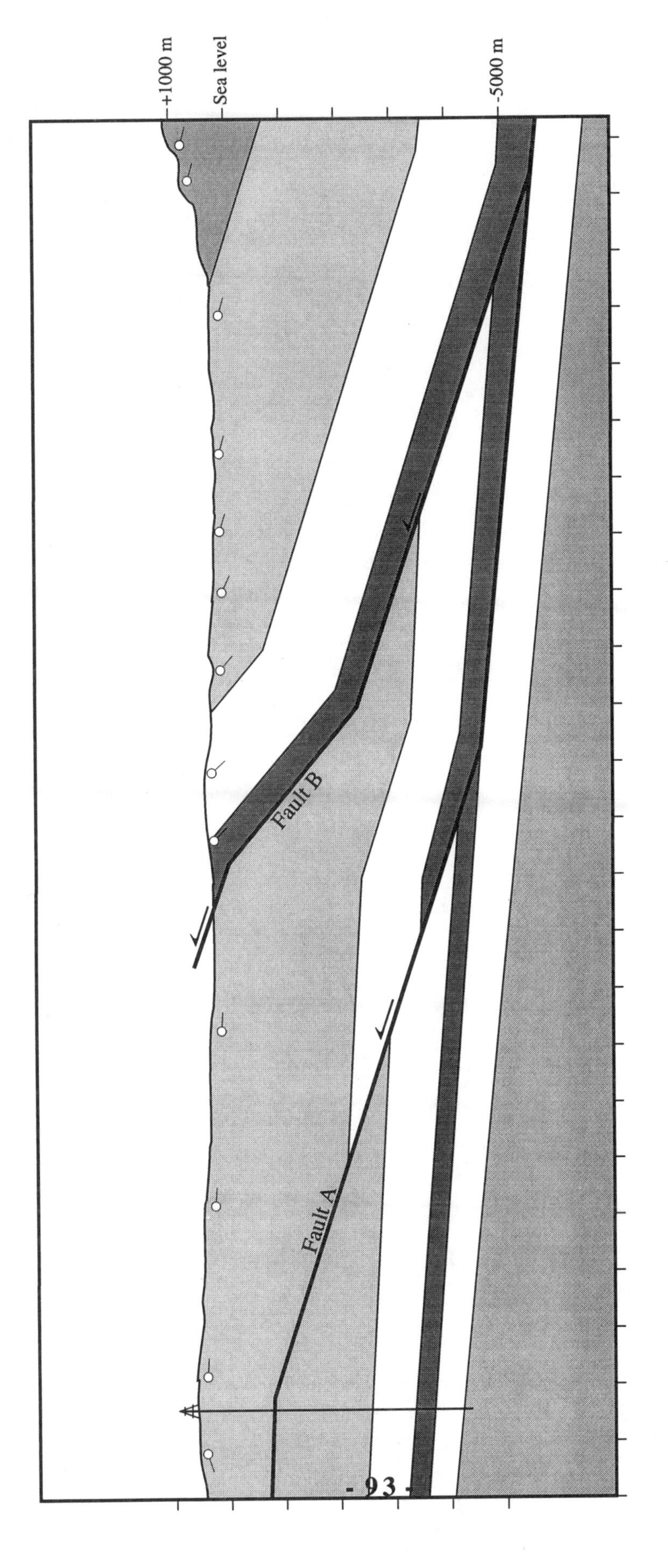

Figure 7.4. Balanced cross section. (After Suppe, 1985)

REGIONAL FOCUS C

THE CASCADIA SUBDUCTION ZONE

INTRODUCTION

Most of the Pacific coast of the United States and Canada is near the boundary between the North American Plate and the Pacific Plate. However, along a 1400 km length of that coast from northern California to southern British Columbia, the diminutive Juan de Fuca Plate (Figure C1) is subducting beneath North America. The effects of subduction are dramatic, including volcanic activity in the Cascade chain, but during historical time, no devastating earthquakes have struck this region.

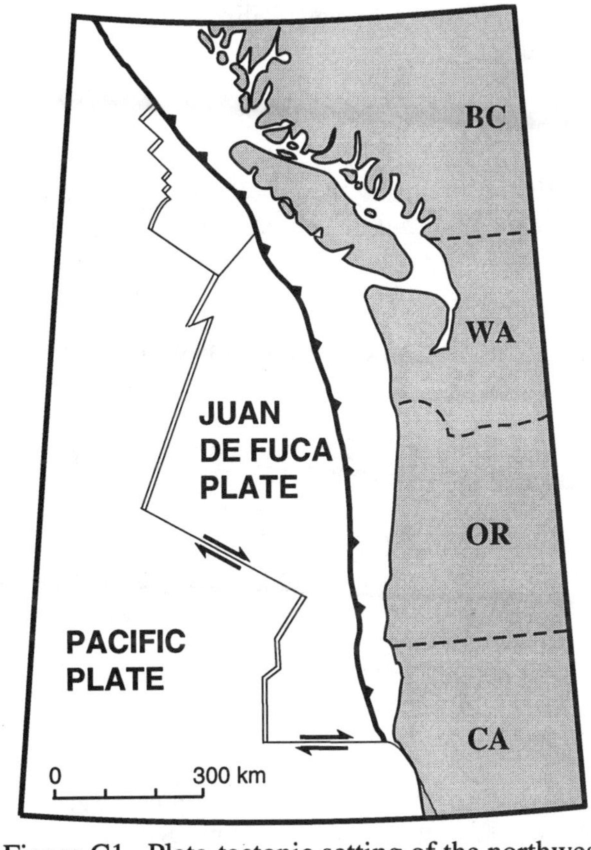

Subduction zones in general are responsible for the most powerful earthquakes on Earth. The thickness of the crust at subduction zones and the slow digestion of the down-going plates can create earthquake ruptures over large areas. The 200-year record of quiet in northern California, Oregon, Washington, and southwestern Canada is a bit of a tectonic mystery. A recent study predicts that a M=6.5 earthquake could cause \$13 to \$26 billion in damage to the Seattle, Washington area alone; a M=8.2-9.3 earthquake (typical of great subduction-zone earthquakes elsewhere) could cause \$51 to \$97 billion in damage.

Figure C1. Plate-tectonic setting of the northwestern United States and southwestern Canada.

The Cascadia Subduction Zone

A growing body of geological evidence suggests that large earthquakes do occur on the Cascadia subduction zone and that the two centuries of historical record fall neatly *between* these earthquakes. Researchers see evidence of surface rupture, severe shaking, and coastal tsunami. Their evidence suggests that the last great earthquake in this area occurred around 1700 AD. Other researchers, however, still dispute these claims. This research has important implications for public safety, construction codes, and disaster preparedness in the Pacific Northwest.

GENERAL REFERENCES

Alper, J., 1993. Shaking Seattle. Earth, 2 (July): 20-25.

Kerr, R.A., 1991. Big squeeze points to a big quake. Science, 252: 28.

Monastersky, R., 1992. Large prehistoric earthquake ripped Seattle. Science News, 142 (Dec. 5): 388-9.

Monastersky, R. Rattling the Northwest. Science News, 137 (Feb. 17): 104-6

Wiens, D.A., 1993. Too hot for earthquakes? Nature, 363: 299-300.

Wuethrich, B., 1994. It's official: quake danger in Northwest rivals California's. Science, 265: 1802-3.

TECHNICAL REFERENCES

Atwater, B.F., 1992. Geologic evidence for earthquakes during the past 2000 years along the Copalis River, southern coastal Washington. Journal of Geophysical Res., 97: 1901-1919.

Clague, J.J.; and P.T. Bobrowsky, 1994. Evidence for a large earthquake and tsunami 100-400 years ago on western Vancouver Island, British Columbia. Quaternary Res., 41: 176-184.

Clarke, S.H. Jr., and G.A. Carver, 1992. Late Holocene tectonics and paleoseismicity, southern Cascadia subduction zone. Science, 255: 188-192.

Darienzo, M.E., C.D. Peterson, and C. Clough, 1994. Stratigraphic evidence for great subduction-zone earthquakes at four estuaries in northern Oregon, USA. Journal of Coastal Research, 10: 850-876.

Kelsey, H.M., and J.G. Bockheim, 1994. Coastal landscape evolution as a function of eustasy and surface uplift rate, Cascadia margin, southern Oregon. Geological Society of America Bulletin, 106: 840-854.

Jacoby, G.C., Jr., P.L. Williams, and B. Buckley, 1992. Tree ring correlation between prehistoric landslides and abrupt tectonic events in Seattle, Washington. Science, 258: 1621-3

Mathewes, R.W., and J.J. Clague, 1994. Detection of large prehistoric earthquakes in the Pacific Northwest by microfossil analysis. Science, 264: 688-691.

McCaffrey, R, and C. Goldfinger, 1995. Forearc deformation and great subduction earthquakes: implications for Cascadia offshore earthquake potential. Science, 267: 856-9.

Mitchell, C.E., P. Vincent, R.J. Weldon II, and M.A. Richards, 1994. Present-day vertical deformation of the Cascadia Margin, Pacific Northwest, United States. Journal of Geophysical Research, 99: 12,257-12,277.

Ng, M.K.F., P.H. LeBlond, and T.S. Murty, 1990. Simulation of tsunamis from great earthquakes on the Cascadia subduction zone. Science, 250: 1248-51.

Various authors, 1992. Five articles dedicated to paleo-earthquakes in Washington state. Science, 258: 1611-1623.

DISCUSSION QUESTIONS

After reading some of the references listed above, you should be prepared to answer the following questions about the Cascadia subduction zone and the potential for earthquakes in the Pacific Northwest:

1) Several researchers see evidence that great subduction-zone earthquakes have repeatedly struck the Pacific Northwest in the recent geological past. Outline the evidence that supports this hypothesis.

2) Other researchers believe that the Cascadia subduction zone is different from other subduction zones and that it is *not* characterized by great earthquakes. Outline the evidence that supports *this* hypothesis.

3) One specific threat that can result from a subduction-zone earthquake is a tsunami. Discuss the evidence of prehistoric tsunami along the Cascadia coastline. Also list two or three alternative explanations for this evidence.

4) How do you think society should deal with the remaining *uncertainty* about the earthquake threat in the Pacific Northwest? Should it assume the worst, or should it avoid widespread alarm?

5) Assume for a moment that scientists could prove conclusively that great earthquakes periodically do occur on the Cascadia subduction zone. What investment in time and money do you think society should make to prepare for the next earthquake? Consider two time scales: 1) the earthquake repeat time in this area (at least 200 years), and 2) human memory time (often less than 200 days). To what extent does the *difference* between these two time scales affect public disaster preparedness?

EXERCISE 8

FAULT-SCARP DEGRADATION

Supplies Needed

- calculator
- straight-edge ruler

PURPOSE

The evolution of the Earth's surface over time is governed by the balance between constructional (tectonic) processes and destructional (erosional) processes. One of the simplest illustrations of this balance is the case of fault scarps, which are steep slopes formed where a fault has ruptured the ground surface during an earthquake. Construction of a fault scarp occurs in seconds and is followed by degradation of the scarp over many years. In some cases, degradation can be modeled numerically, so that the shape of the slope in the past or in the future can be inferred. This tool is particularly valuable in paleoseismology because it sometimes can be used to determine the ages of ancient earthquakes from fault scarps that otherwise would be impossible to date. This exercise will show you the basics of slope degradation modeling and how to use such models to date fault scarps.

INTRODUCTION

How slopes evolve over time is a question that geomorphologists have been debating for more than a century. William Morris Davis (1899) argued that erosion over long periods of time tends to make slopes less steep and more rounded. In contrast, Walther Penck (1924), a German geomorphologist, argued that slopes represent a dynamic balance between erosion and uplift. According to Penck, slopes were straight, convex, or concave as a result of changes in uplift rates as the slopes formed. The followers of Davis and Penck waged ideological warfare for many decades until most scientists realized that ideology makes bad science. Davis' theory underestimated the variability of tectonic processes, and Penck's theory underestimated the variability of erosional processes. Most geomorphologists today focus on the individual processes that shape the landscape.

Weathering-limited slopes vs. Transport-limited slopes

Two steps are necessary for erosion to act upon a surface: weathering and transport. Weathering is the step that loosens rock. Various physical, chemical, and biological processes act to pull rock apart into individual blocks, grains, or molecules. Those processes alone, however, are insufficient to erode anything. Where weathering alone occurs on a surface, a cover of loose sediment or soil accumulates, eventually covering the rock and protecting it from further weathering action. For erosion to occur, weathered material must be *transported* away.

Both weathering and transport are present on most slopes, but one of the two usually occurs somewhat faster than the other. Which process is faster profoundly shapes a slope. The slower process *limits* erosion. Geomorphologists classify slopes as either **weathering limited** (weathering occurs more slowly than transport) or **transport limited** (transport is the slower process). Weathering-limited slopes are characterized by bare rock exposed at the surface and gradients that can be very steep, even vertical. Transport-limited slopes are characterized by a cover of sediment at the surface. Unconsolidated sediment can never be steeper than the angle of repose of the material (25-30° in sand), so that transport-limited slopes tend to be less steep. The dominant erosional process on transport-limited slopes is **creep**, which refers to the slow movement of sediment as a result of gravity. Other processes such as raindrop impacts and frost action also contribute to downslope creep. Most importantly, sediment creep is a process that can be simplified and represented by a numerical model. In contrast, erosion on weathering-limited slopes is much more difficult to quantify.

Diffusion Modeling

The evolution of a transport-limited slope through time can be evaluated quantitatively by assuming that sediment transport is a **diffusion process**. Diffusion also describes how chemicals in solution move from areas of high concentration to areas of low concentration, and how heat moves from areas of high temperature to areas of low temperature. In sediment diffusion, gravity carries sediment from an area of high elevation (the top of the slope) towards an area of low elevation (the base of the slope). Without fresh uplift or downcutting, diffusion tends to make slopes smoother and less steep over time.

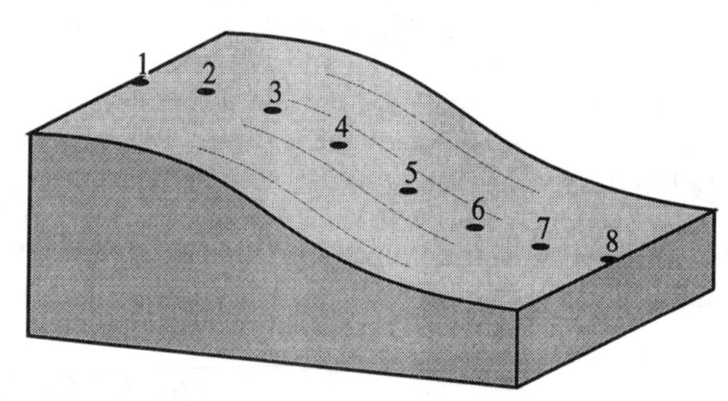

Figure 8.1. An elevation profile across a slope is measured at several points.

In the field, a geologist typically measures the elevations of a series of points in a line perpendicular to the scarp using surveying equipment. The heights of those points and the distances between them are used to construct a cross section.

In order to calculate rates of sediment diffusion, it is necessary to subdivide a slope into a series of short segments. Calculations are easiest if all segments have the same width. If a slope has been profiled in the field, then slope segments can be the intervals between measured points. Diffusion models look at each of these segments as a tall column of sediment (Figure 8.2). Like a series of small basins over which a waterfall flows, each segment receives input (water on the waterfall; sediment on the slope) from the segment above it and sends discharge into the segment below.

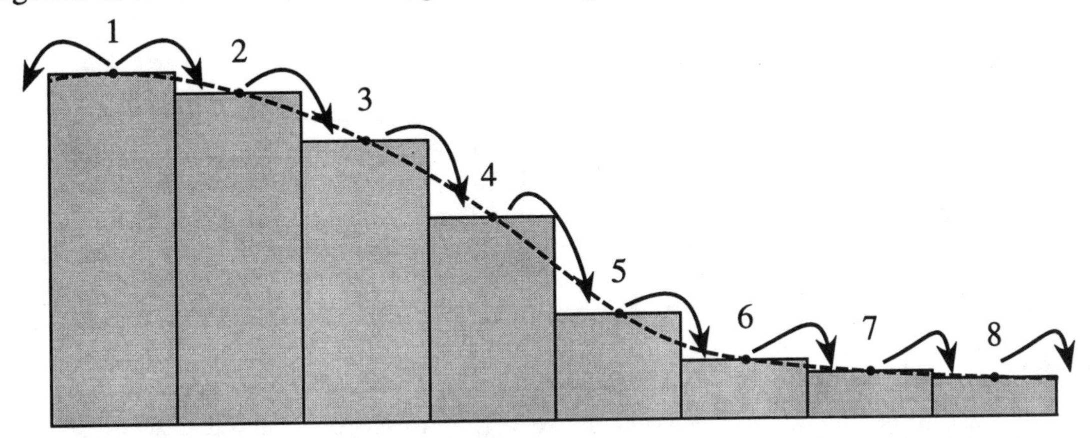

Figure 8.2. In order to model a slope, it is convenient to imagine the slope as several discrete segments with sediment moving from the highest segment into successively lower segments.

Sediment diffusion can be evaluated quantitatively by using the **Continuity Equation**:

$$\frac{\delta z}{\delta t} = \frac{\delta R}{\delta x} \qquad (8.1)$$

where **z** is the height (elevation) at any given point an a slope, **R** is the rate of sediment movement, **x** is horizontal distance, and **t** is time (Table 8.1 below summarizes all the parameters used in this chapter). The Continuity Equation is one of the simplest equations in science – translated into English, it simply states that the change in elevation through time at a point equals the difference between the amount of sediment *arriving at* that point and the amount of sediment *leaving*. For example, a 10 m wide slope segment that receives 20 m^2 of sediment from upslope (on a two-dimensional cross section, volume is measured in m^2) and sends 30 m^2 downslope thereby loses 10 m^2, and an average of 1 m of material erodes from that 10 m length of the slope.

Table 8.1. Parameters used in calculating slope diffusion.

Parameter	Explanation	Units
z	elevation	meters
t	time	years
R	sediment flux rate	m^2/ yr
x	horizontal position	meters
$\delta z/\delta t$	elevation change over time	m / yr
$\delta R/\delta x$	change in transport rate	m^2/ yr^2
κ	diffusivity	m^2/ yr
$\delta z/\delta x$	slope gradient	none

Fault-Scarp Degradation

There is a second equation on which diffusion modeling is based. This equation relates the rate of sediment transport (R), a constant (κ, called *diffusivity)*, and the slope gradient ($\delta z/\delta x$):

$$R = \kappa \times \frac{\partial z}{\partial x.} \tag{8.2}$$

Other, more complicated versions of this equation exist, but this simpler version can be used where sediment transport is caused mainly by creep. The diffusivity constant (κ) is very important here. Diffusivity characterizes how easily the sediment can be moved and how much the local climate can move it. Diffusivity varies substantially from one area to another. In order to estimate the sediment-transport rate (R), it is necessary to measure diffusivity in the field or infer it indirectly.

Example 8.1: Modeling Slope Evolution

This example illustrates how you can use a slope-diffusion model to determine how a given slope will change over time. This example lists the steps necessary to predict how the shape of the slope shown below will change after a 1000 year jump in time. Assume that the constant of diffusivity (κ) has been measured in the field and is about 50 m^2/yr.

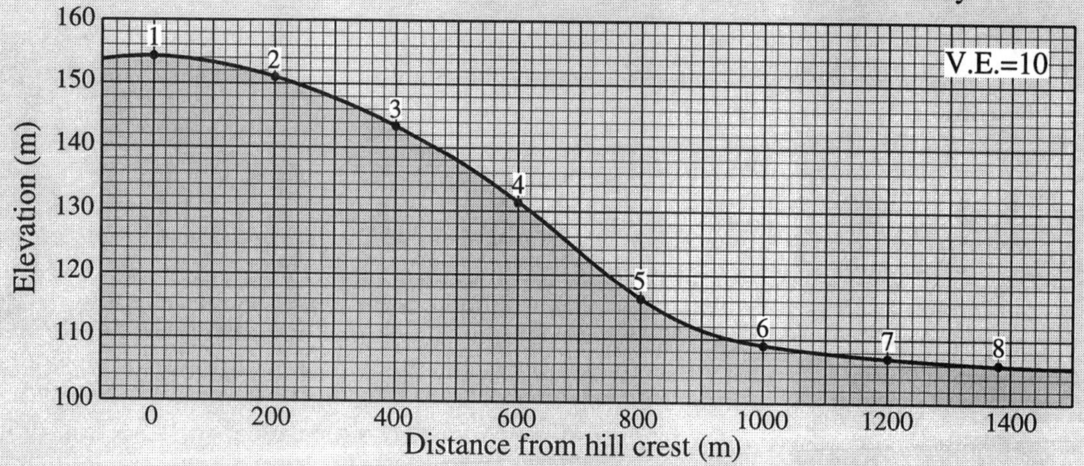

In order to model this slope using sediment diffusion, we subdivide the profile shown above into eight points, each one separated by a distance of 200 m. The coordinates (in meters) of each point are given below:

	1	2	3	4	5	6	7	8
Distance (m) = x =	0	200	400	600	800	1000	1200	1400
Elevation (m) = z =	154.0	151.0	144.0	131.0	116.0	109.0	107.0	106.0

Equation 8.1 shows that the rate of sediment transport between any two adjacent points on a slope is a function of how steep the slope is between those points. The average gradient between two points is the change in elevation (δz) divided by the change in distance (δx). In this example, δx is always 200 m. The gradient is dimensionless (rise ÷ run) and is not expressed as an angle. For example, the average gradient between Point #1 and Point #2 is (154.0 m – 151.0 m) ÷ (200 m – 0 m), or 0.015.

Gradient = $\frac{\delta z}{\delta x}$	1–2	2–3	3–4	4–5	5–6	6–7	7–8
	0.015	0.035	0.065	0.075	0.035	0.010	0.005

The constant of diffusivity (κ) in this example is 50 m^2/yr, and Equation 8.2 tells you that the rate of sediment transport (R) between a pair of points is the product of diffusivity and the gradient:

Sed. transport rate = R = (m^2/yr)	1–2	2–3	3–4	4–5	5–6	6–7	7–8
	0.75	1.75	3.25	3.75	1.75	0.5	0.25

The rate of sediment transport (R) multiplied by time (1000 years in this example) gives the amount of sediment (in m^2) that moves from the upslope point to the downslope point during that time.

Sed. flux = R∗1000 yrs = (m^2)	1–2	2–3	3–4	4–5	5–6	6–7	7–8
	750	1750	3250	3750	1750	500	250

Equation 8.1 tells you that the change in elevation of any point on a slope is the difference between the amount of sediment transported to that point and the amount of sediment transported away. For example, Point #3 receives 1750 m^2 during this 1000 year interval but loses 3250 m^2 downslope. The net loss of 1500 m^2 of sediment averaged across the 200 m length of the slope in that segment corresponds to an average of 7.5 m of erosion.

There are two exceptions to this rule. At Point #1 (at the crest of the hill), sediment moves downslope in *two* directions. As a result, that point erodes at *twice* the rate calculated from transport to the right alone. At Point #8 (at the base of the slope), you should assume that the elevation does not change because you have no information about the slope below that point. The boundaries of models like this one are often awkward.

	1	2	3	4	5	6	7
Sed. gained - sed. lost = (m^2)	- 1500*	- 1000	- 1500	- 500	+2000	+1250	+250
Elevation change (m) =	-7.5	-5.0	-7.5	-2.5	+10.0	+6.25	+1.25

The elevation of each point after the 1000-year interval is the sum of the old elevation and the elevation change caused by sediment diffusion.

	1	2	3	4	5	6	7	8
NEW elevation (m) =	146.5	146.0	136.5	128.5	126.0	115.25	108.25	106.0

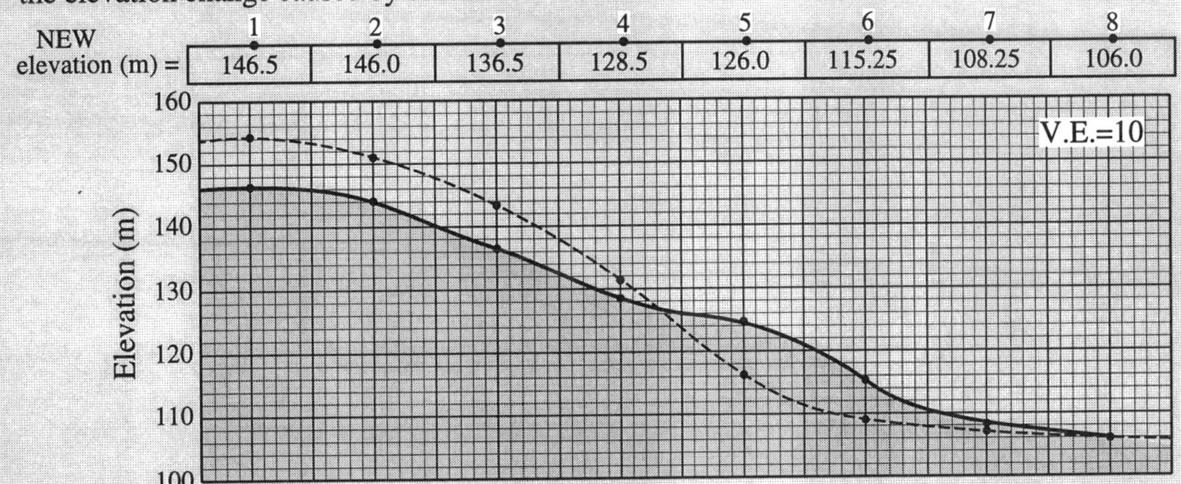

Fault-Scarp Degradation

In the example on the preceding pages, the slope after 1000 years is somewhat irregular because the example subdivides the slope into only eight points and because the 1000 years were lumped into a single interval. Diffusion calculations are done easily on computers, and a computer model could easily subdivide the example above into thousands of points and thousands of small time steps.

Now, follow the steps outlined in Example 8.1 in order to solve the problem below. In this problem, you will determine how the slope shown changes through four 100-year time steps. Use a constant of diffusivity of $\kappa = 100\ m^2/yr$. After each time step, plot the new profile on the graph. Keep the elevations of both Point #1 and Point #8 constant (i.e., assume that both points receive as much sediment as they lose).

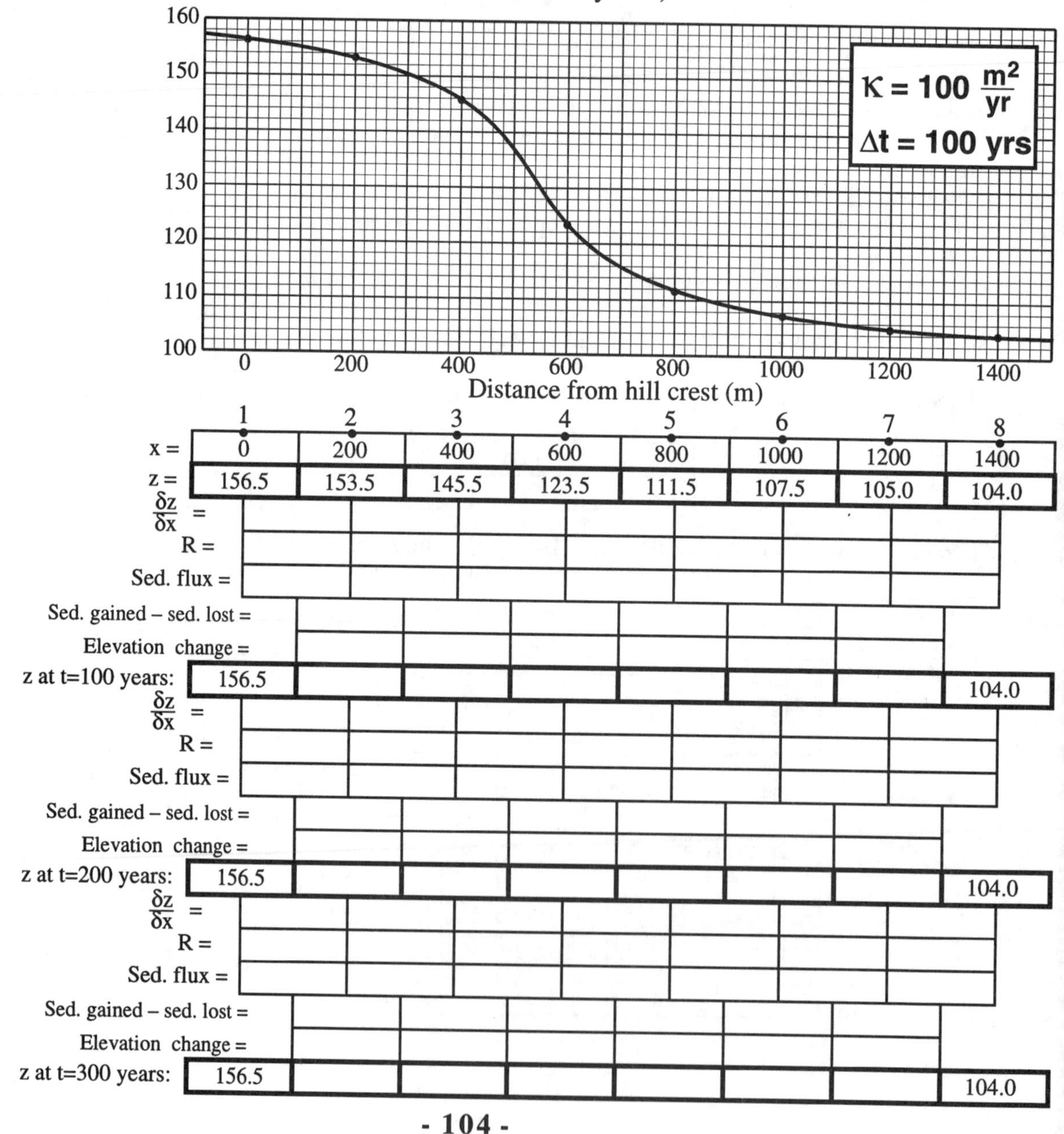

	1	2	3	4	5	6	7	8
x =	0	200	400	600	800	1000	1200	1400
z =	156.5	153.5	145.5	123.5	111.5	107.5	105.0	104.0
$\frac{\delta z}{\delta x}$ =								
R =								
Sed. flux =								
Sed. gained – sed. lost =								
Elevation change =								
z at t=100 years:	156.5							104.0
$\frac{\delta z}{\delta x}$ =								
R =								
Sed. flux =								
Sed. gained – sed. lost =								
Elevation change =								
z at t=200 years:	156.5							104.0
$\frac{\delta z}{\delta x}$ =								
R =								
Sed. flux =								
Sed. gained – sed. lost =								
Elevation change =								
z at t=300 years:	156.5							104.0

FAULT-SCARP DEGRADATION

A *scarp* is defined as a steep slope or a steep portion of a larger, less steep slope. Scarps can form as a result of many different geomorphic processes. *Fault scarps* form when a fault ruptures the surface during an earthquake. Quantitative modeling of sediment diffusion has proven extremely useful when it is applied to fault scarps. Specifically, when a geologist identifies a fault scarp in the field, he or she most commonly wants to find out *when* the earthquake that caused that fault scarp occurred. Given the shape of the fault-scarp profile, given assumptions about how the scarp looked when it first formed, and given the value of the diffusivity constant, the geologist can estimate when the scarp formed.

Fault scarps that cut unconsolidated sediment or soil are very promising for diffusion modeling because they form instantaneously and then systematically degrade afterwards. Older fault scarps are smoother and less steep than recent fault scarps. In principle, a geologist could measure a profile across any fault scarp and then calculate exactly how much time is represented by the degradation of the profile. In practice, several criteria must be met for a solution to be possible:

A) The scarp must be transport-limited. Fault scarps on bedrock cannot be modeled using diffusion.

B) After the earthquake occurred, the scarp must have quickly collapsed to the angle of repose (25-30° in sand).

C) It must be possible to measure, infer, or assume a value for diffusivity (κ).

D) The scarp must have formed in a single rupture event.

Of these criteria, the third often is the most difficult to meet. It's never possible to determine the age of a fault scarp without knowing the value of diffusivity. One method for inferring diffusivity is outlined in the exercise that follows.

Given the four criteria above, several solutions give the time (t) since a fault scarp formed. The following is the solution given by Colman and Watson (1983):

$$\kappa t = \frac{d^2}{4\pi} \frac{1}{(\tan \theta - \tan \alpha)}, \qquad (8.3)$$

where d is the vertical separation between the upper slope and the lower slope, θ is maximum scarp angle, and α is far-field slope angle (see Table 8.2). These parameters can be measured easily on a cross section across a fault scarp, as illustrated in Figure 8.3.

Table 8.2. Additional parameters for calculating fault-scarp degradation (also see Table 8.1 and Figure 8.3).

Parameter	Explanation	Units
d	vertical displacement on a scarp	meters
π	pi = 3.14159	none
θ	maximum scarp slope angle	degrees
α	average far-field slope angle	degrees

Fault-Scarp Degradation

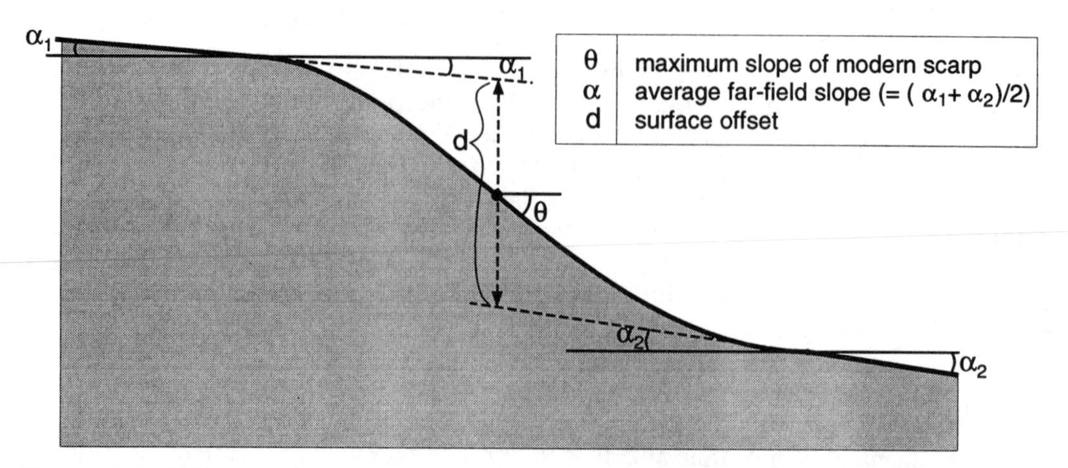

Figure 8.3. Measurements that need to be made on a fault scarp in order to solve Equation 8.3.

1) Using the elevation and distance measurements for the ten points in profile A-A', plot a cross section of the fault scarp.

Point #	Dist. (m)	Height (m)
1	0	0.00
2	10	0.30
3	20	0.90
4	30	2.00
5	40	4.20
6	50	7.50
7	60	8.70
8	70	9.80
9	80	10.40
10	90	10.60

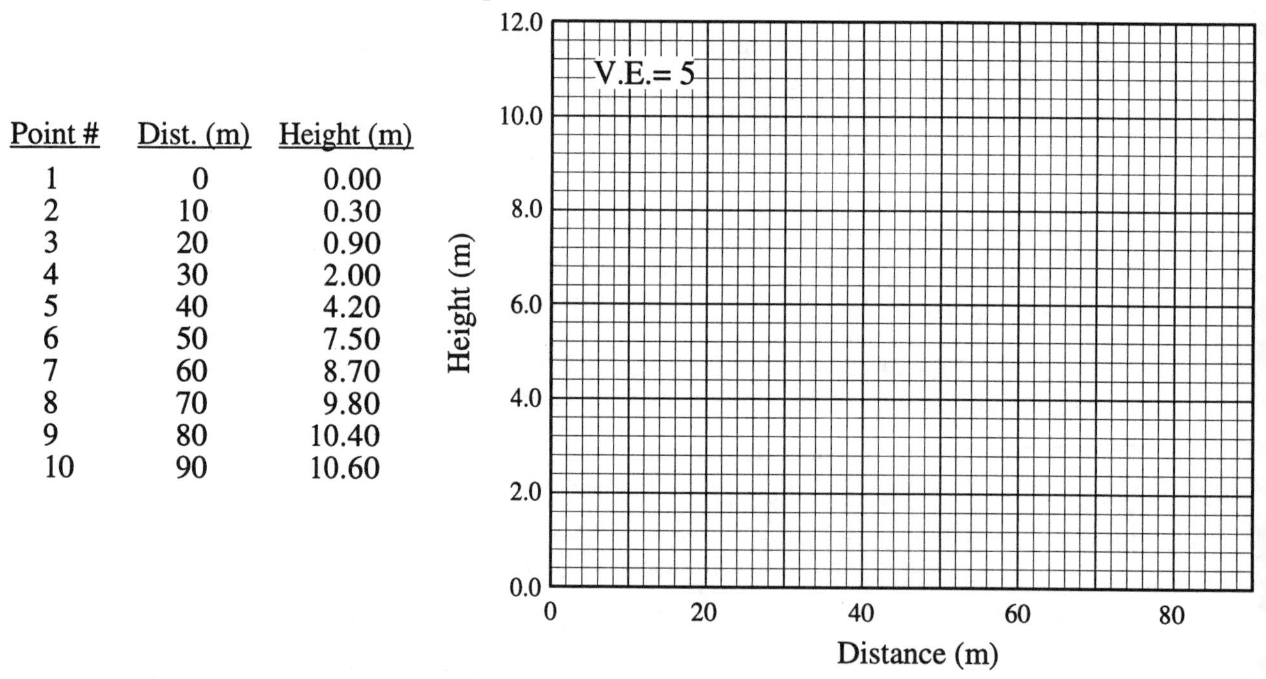

2) Calculate the gradient angle between points #1 and #2 and the gradient angle between Points #9 and #10 (remember that a gradient angle equals arctan[rise ÷run]). You should see that these two angles are α_1 and α_2. Calculate α for this scarp.

3) Find the steepest interval between adjacent points on this profile. Calculate the gradient angle of that interval. This angle is θ.

4) Using the graph on the above, measure d for this scarp. Remember to convert your measurement (cm on this page) into real-world height (in meters) using the vertical scale on the graph.

5) Assume that this fault formed 120 years ago (t=120 yrs). Use Equation 8.3 to calculate the value of diffusivity on this scarp.

Equation 8.3 and other solutions like it assume that the faulting event (at t=0) cut a scarp across a smoothly sloping surface made of homogeneous and unconsolidated sediments, and that that scarp degraded to the angle of repose immediately. This solution provides a simple and easily applicable relationship between the geometry of a scarp today, diffusivity, and the age of the scarp.

Fault-Scarp Degradation

The Two-Scarp Problem

A geologist has identified two fault scarps along an active fault zone. Historical records show that one of the scarps (Scarp A) formed during a damaging earthquake 200 years ago. The other scarp (Scarp B) formed during some unknown earthquake during prehistoric time.

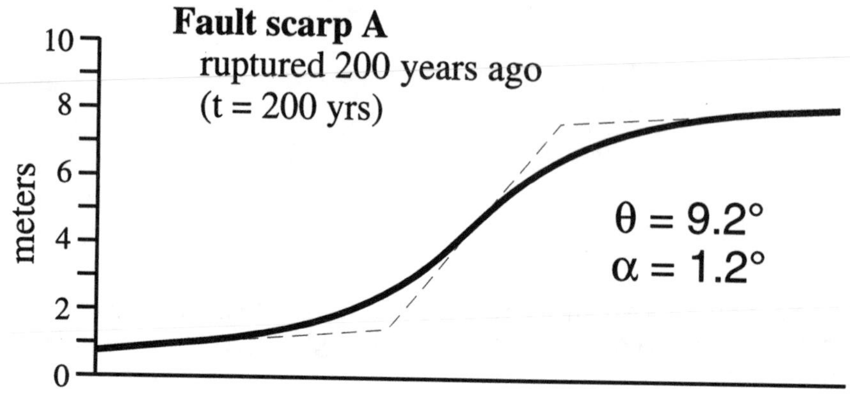

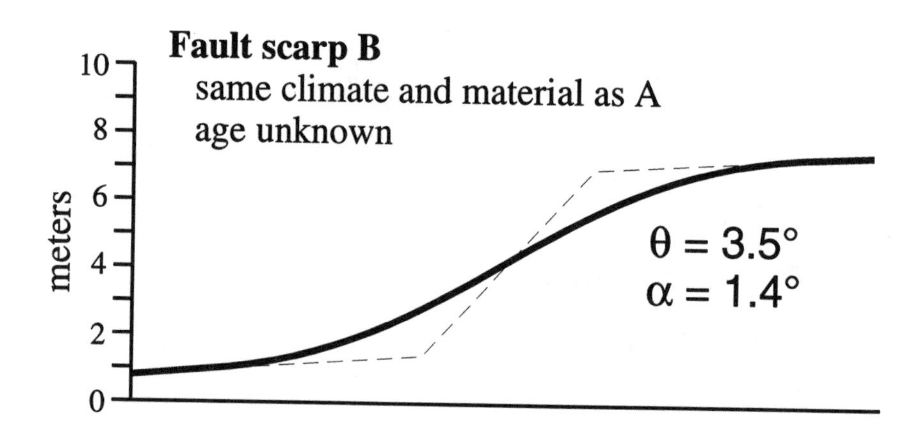

6) Using Scarp A, calculate the value of diffusivity (κ).

7) Assuming that the two scarps are cut in the same material, estimate the age of the earthquake that formed Scarp B.

8) Given only the information in this two-scarp problem, estimate the average recurrence interval of large earthquakes on this fault zone.

9) Predict in what year the next ground-rupturing earthquake will occur on this fault zone. What are some problems with this prediction?

BIBLIOGRAPHY

Andrews, D.J., and R.C. Bucknam, 1987. Fitting degradation of shoreline scarps by a nonlinear diffusion model. Journal of Geophysical Research, 92: 12,857-12,867.

Bucknam, R.C., and R.E. Anderson, 1979. Estimation of fault-scarp ages from a scarp-height-slope-angle relationship. Geology, 7: 11-14.

Colman, S.M., and K. Watson, 1983. Ages estimated from a diffusion equation model for scarp degradation. Science, 221: 263-265.

Davis, W.M., 1899. The geographical cycle. Geographical Journal, 14: 481-504.

Hanks, T.C., and D.J. Andrews, 1989. Effects of far-field slope on morphological dating of scarplike landforms. Journal of Geophysical Research, 94: 565-573.

Nash, D.B., 1986. Morphological dating and modeling degradation of fault scarps. In R.E. Wallace, ed., Active Tectonics. National Academy Press, Washington D.C.

Nash, D.B., 1980. Morphological dating of degraded normal fault scarps. Journal of Geology, 88: 353-360.

Penck, W., 1924. Morphological Analysis of Landforms (translated by Ezech and Boswell in 1953). MacMillian, London, 429 pp.

Quocheng, S., and C. Yin-Mou, 1992. Applications of the diffusion model in slope evolution - some case studies in Taiwan. Journal - Geological Society of China, 35: 407-419.

Stewart, I.S., 1993. Sensitivity of fault-generated scarps as indicators of active tectonism: some constraints from the Aegean region. In D.S.G. Thomas and R.J. Allison (eds.), Landscape Sensitivity. John Wiley & Sons: New York.

EXERCISE 9

FAULT TRENCHING

Supplies Needed

- calculator
- straight-edge ruler

PURPOSE

Paleoseismology is the study of earthquakes in the recent geologic past. In particular, the focus of research is on the occurrence, size, and timing of prehistoric earthquakes. The history of past earthquakes is the best tool for predicting the location, size, and frequency of future earthquakes. Past earthquakes are recorded best where faulting coincides with active deposition. Fault trenching is a method for exposing faulted sediments and deciphering the history of earthquake activity. The purpose of this exercise is to familiarize you with the technique of fault trenching, illustrate stratigraphic evidence of earthquakes, and show you how to interpret that evidence.

INTRODUCTION

The principle challenge in many fields of geology is seeing the features to be studied. Field geologists scour an area for useful outcrops and interpolate their data between these fixed points; petroleum geologists drill kilometers through sedimentary rock to infer the stratigraphic and structural relationships beneath the surface; and seismologists use seismic waves to interpret the composition of the Earth's core and mantle. Geologists who study active faults are at the same disadvantage – earthquakes that occur today may rupture the surface, but erosion quickly erases that evidence in many settings. The history of past earthquakes is the key to predicting future earthquakes, but that history often lies buried beneath the surface.

Fault trenching is a technique developed in the last couple of decades to reveal evidence of past earthquakes in near-surface sediments. The principle is that if the fault won't come up to the geologist, the geologist will go down to the fault. Using a bulldozer, backhoe, or pick and shovel, one or more long ditches are cut across a fault that is active or suspected to be active. A scientific fault trench is not just a hole in the ground, but should have the following features:

Fault Trenching

- the trench should cross a fault or suspected fault
- one or more of the trench walls should be vertical
- the trench should cut sediments that accumulated during the period of fault activity
- the sediments should contain material suitable for radiocarbon or other numerical dating techniques.

Trench walls are vertical because only vertical walls clearly display the detailed stratigraphic layers and fault structures that need to be studied. The most successfully trenches reveal a detailed stratigraphy that accumulated during the same time interval that the fault was active. A Cambrian sedimentary rock that accumulated between 610 and 600 million years ago, for example, would provide little information about a fault active in the last few thousand years. Ideally, the stratigraphy in a fault trench consists of fine layers rich in organic material, so that the ages of earthquake events on that fault can be tightly bracketed by numerical ages.

Determining the ages of earthquake events revealed in a fault trench relies upon the two most basic axioms of stratigraphy: the principle of *superposition* and the principle of *cross-cutting relationships.* The principle of superposition states that sedimentary layers accumulate from the bottom up, so that any layer is younger than the layers beneath it and older than the layers above (Figure 9.1A). The principle of cross-cutting relationships states that if geological feature #1 cuts geological feature #2, then feature #2 is older than feature #1. In Figure 9.1B below, fault Y cuts layer C. We therefore know that layer C was deposited before Y last ruptured.

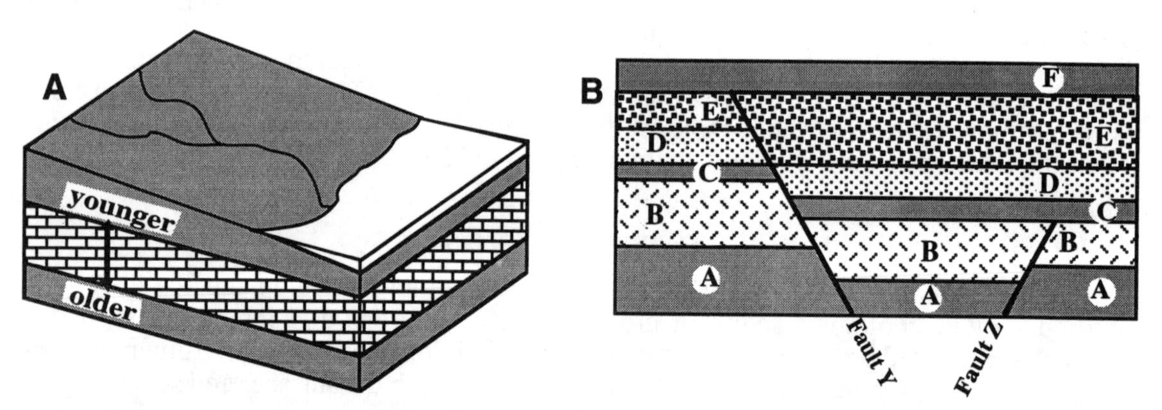

Figure 9.1. The principle of superposition (A) and the principle of cross-cutting relationships (B).

The location of a fault trench or a series of trenches is crucial to the success of the investigation. Most importantly, the location of the fault should be determined with fair precision before any excavation begins. The geologist almost certainly does not own the land he or she is tearing up, and disruption of the surface should be as little as possible to accomplish the scientific goals. In addition, excavating equipment often costs $100 per hour or more, so good planning is essential. When a geologist digs part of a trench by hand, the punishment for poor planning is lost days, sore muscles, blistered hands, and untold frustration.

One of the groundbreaking studies (so to speak) that utilized fault trenching was an investigation of the San Andreas fault at Pallett Creek, about 55 km (35 mi) northeast of Los Angeles, California (Sieh, 1978; Sieh, 1984). The segment of the San Andreas ruptured last in 1857 in an earthquake that was at least M_w=7.8-7.9, with slip on the fault between 3 and 4.5 m (10-15 ft). The age and displacement of the 1857 earthquake was known, but geologists had no history of previous earthquakes from which to calculate future earthquake hazard. Pallett Creek was a promising site because the area was swampy until the early 1900s, after which the creek began incising, cutting a 10 m deep canyon and lowering the groundwater table in the surrounding sediments. Figure 9.2 shows the series of trench cuts made on the north side of Pallett Creek to expose sediments cut by the San Andreas fault. The different exposures were cut using a backhoe (Exposures 10, 10a, 11, 11a, and 11b), a bulldozer (Exposure 5), and by hand (Exposures 1, 2, 3, 5, and 7).

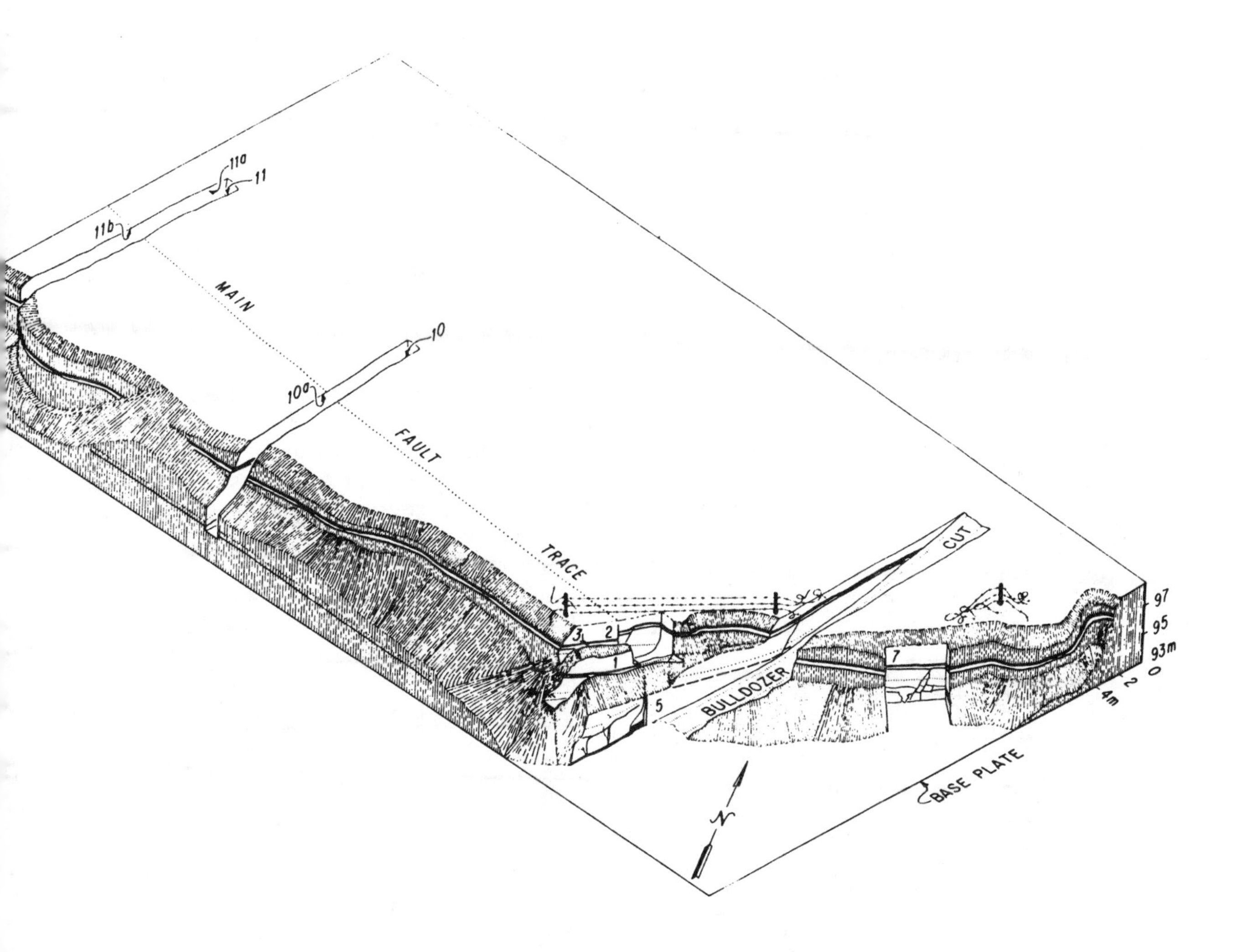

Figure 9.2. Excavations across the San Andreas fault at Pallett Creek, about 55 km northeast of Los Angeles

INTERPRETING EARTHQUAKE EVENTS IN A TRENCH EXPOSURE

Earthquake events can be recorded in a stratigraphic sequence by a number of different features. The clearest evidence of fault rupture is *displaced strata* (Figure 9.3A). Where strata are continuous over distances of a few meters or more, distinct fault breaks may be recognized. The highest layer that is broken by a given fault strand gives the *maximum* age of that earthquake event; the lowest layer that is not broken gives the *minimum* age. Note that the amount of offset on pure normal and reverse faults can be measured directly from displaced strata, but displacement does not give the offset on strike-slip or oblique faults.

Where a fault ruptures in or near wet and unconsolidated sediments, *sand boils* often form (Figure 9.3B). These features, also called "sand craters" or "sand volcanoes," form when seismic shaking liquefies sediment beneath the surface. Sand is less dense than silt and clay, so that liquefied sand is forced to the surface by the weight of the overlying material. After the 1811-12 earthquake sequence near New Madrid, Missouri (see Regional Focus D), numerous sand boils were found on the surface in the epicentral area. In cross section, a sand boil consists of a vertical feeder pipe connected to a filled crater and a sheet of sand or silt that covered the surface at the time of the earthquake.

Intense seismic shaking also can crack the surface, both near a fault and some distance away. Continued deposition of sediment fills these cracks. In a stratigraphic sequence near an active fault, these *filled fissures* (Figure 9.3C) can indicate an earthquake event if several of them occur at the same horizon.

The fourth stratigraphic feature that may indicate ancient earthquakes is a *colluvial wedge* (Figure 9.3D). Colluvium is sediment deposited by gravity. When a fault scarp forms in unconsolidated sediment, it often is steeper than the angle of repose of that material. Gravity carried the excess sediment down the slope, forming a wedge-shaped mass of colluvium at the base of the scarp. Continued deposition at the site buries the colluvial wedge. In a trench, colluvial wedges can be recognized by their shape, poor sorting of the material, and by intimate association with a fault plane.

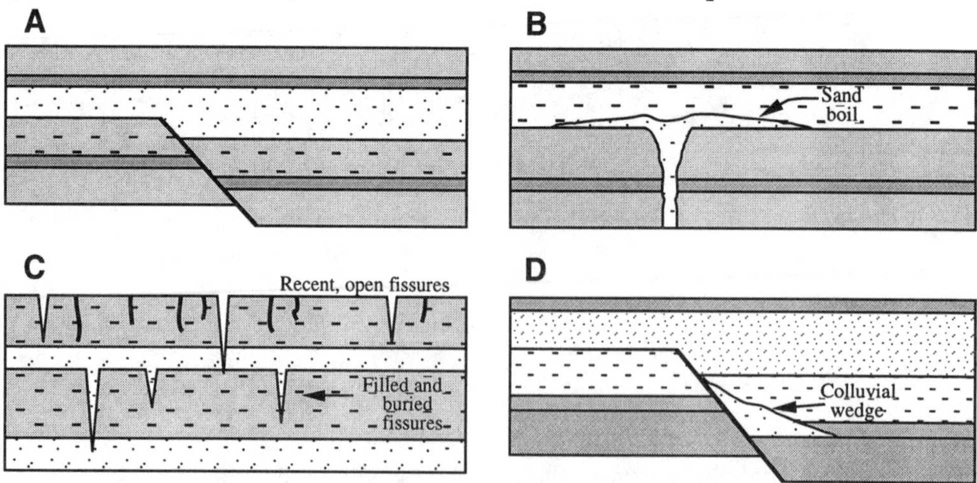

Figure 9.3. Stratigraphic evidence of earthquakes.

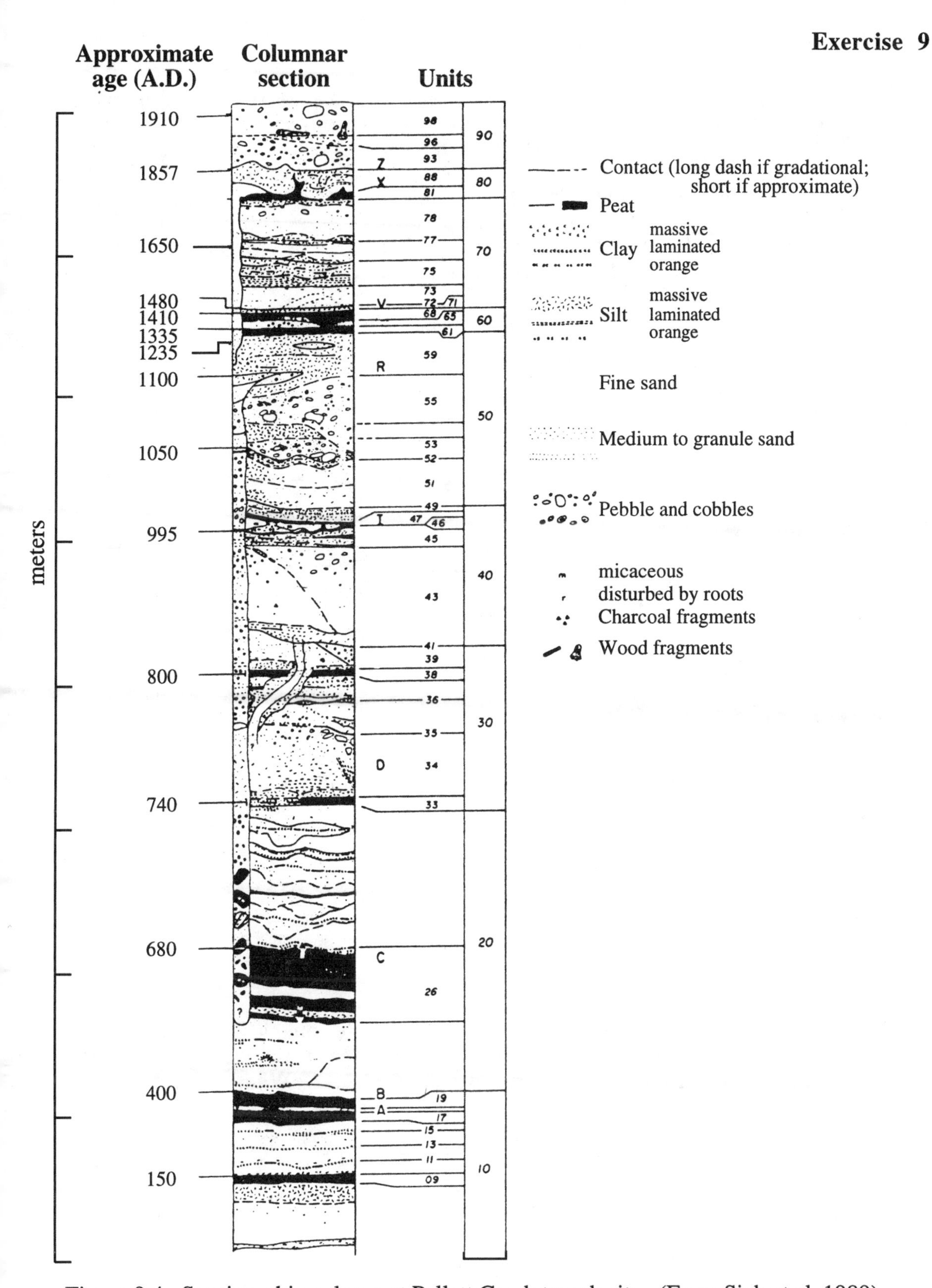

Figure 9.4. Stratigraphic column at Pallett Creek trench site. (From Sieh et al, 1989).

Fault Trenching

Figure 9.4 on the previous page is a generalized description of the stratigraphy at Pallett Creek, including the results of radiocarbon dates from several of the different layers. The ages of layers without radiocarbon dates can be bracketed using the principle of superposition. For example, layer 55 must have been deposited after 1050 A.D. but before 1100 A.D. You will use these dates to determine the ages of some of the earthquake events on the San Andreas fault at this site. The principle of cross-cutting relationships is the most useful tool for determining the age of an earthquake. For example, if layer 59 is displaced by a fault rupture but layer 61 is not cut, then that rupture must have occurred after 1100 A.D. but before 1335 A.D.

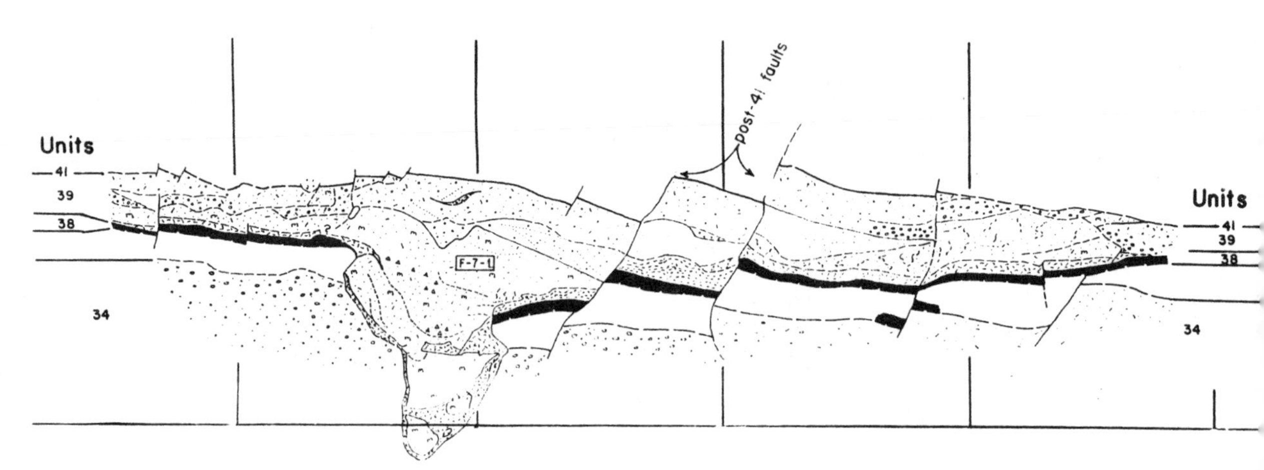

Figure 9.5. Exposure 7 at Pallett Creek.

The figure above (Figure 9.5) illustrates at least two different earthquake events. Locate the area labeled F-7-1 in Figure 9.5.

1) What kind of feature is F-7-1?

2) Assume that the numbers on the left side of Figure 9.4 are numerical ages that you can use to bracket different strata as well as earthquake events. For example, stratigraphic unit 13 formed between 150 and 400 A.D. What are the maximum and minimum ages that bracket the earthquake that formed F-7-1?

3) Enter that age information into Table 9.1 on the next page for Earthquake Event F.

4) Examine Figure 9.6. Determine the maximum and minimum ages for Earthquake Events N and T. Enter this information in Table 9.1.

5) Complete Table 9.1. To estimate the actual date of each earthquake event, you will average the maximum and minimum estimates. For example, Earthquake Event X occurred some time between 1753 and 1817 A.D. Using this method, you'll estimate that the earthquake occurred around 1785 A.D. In addition, you will estimate the recurrence interval between each pair of earthquakes. For example the interval between Event X (1785 A.D.) and Event Z (1857 A.D.) is 72 years.

6) Estimate the *average recurrence interval* for earthquakes on this segment of the San Andreas fault by averaging the times between all of the earthquakes recorded at Pallett Creek. Enter this estimate at the bottom of Table 9.1.

Table 9.1. Earthquake events at Pallett Creek. (Dates from Sieh et al, 1989)

Earthquake Event	Overlying unit / Underlying unit	min. age (A.D.) / max. age (A.D.)	average age estimate ([min.+ max.]÷2)	Recurrence Interval (yrs)
Z	93 / 88	Jan. 9, 1857	Jan. 9, 1857	
				72
X	88 / 81	1817 / 1753	1785	
				305
V	71 / 68	1495 / 1465	1480	
				________
T	________	________	________	
				________
R	61 / 55+	1165 / 1035	________	
				________
N	________	________	________	
				________
I	47 / 45	1013 / 981	________	
				________
F	________	________	________	
				________
D	35 / 34	747 / 721	________	
				________
C	27 / 25	684 / 658	________	
				________
B	20 / 18	400 / 210	________	

Average Recurrence Interval = ________

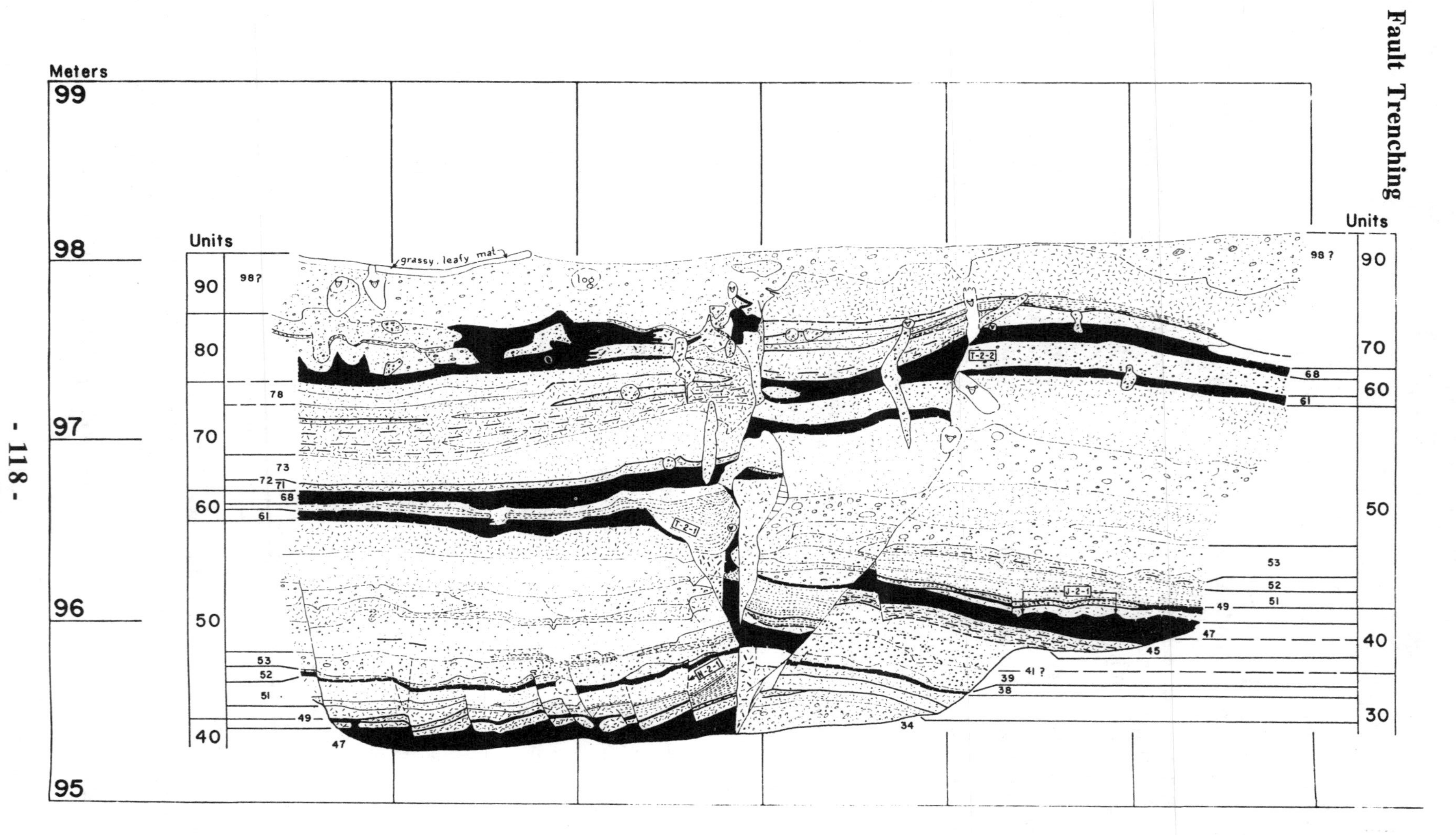

Figure 8.6. Exposure 2 at Pallett Creek.

7) Plot the ages of all of the earthquakes in Table 9.1 in the graph below. Because each age is merely an estimated range, you will need to draw each earthquake as a horizontal bar. Earthquake Event B is plotted for you as an example.

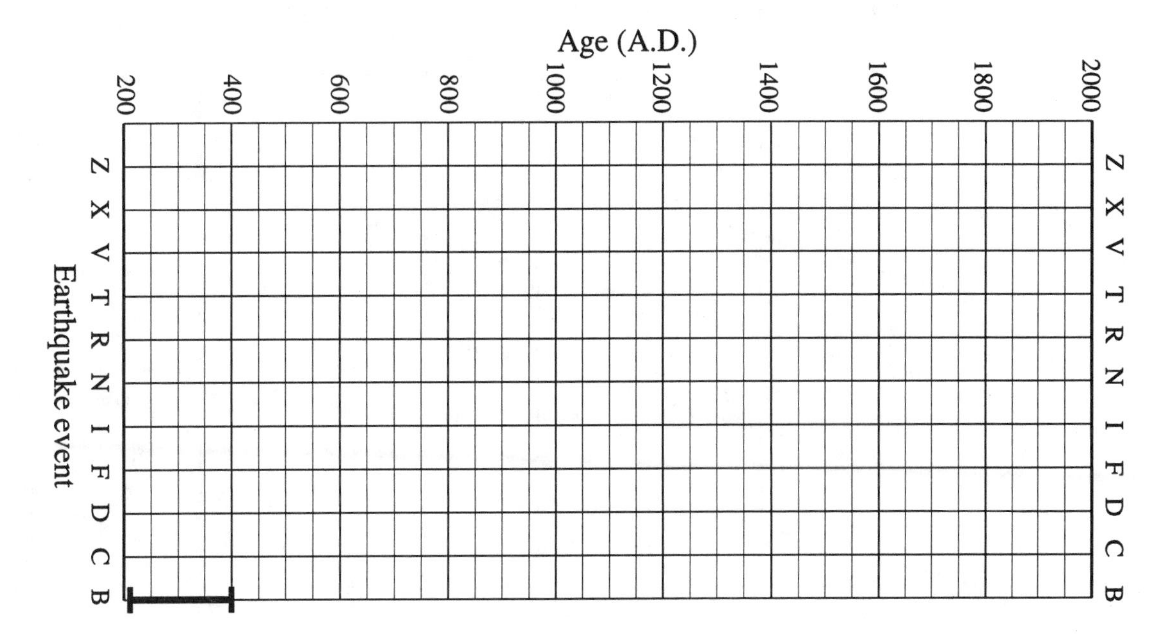

8) Can you connect all or most of the earthquakes in the graph above with a straight line? The *characteristic earthquake* model says that the same faults are characterized by earthquakes with roughly the same magnitude that occur at roughly equal time intervals. How well does this model seem to work for this segment of the San Andreas fault?

An additional piece of evidence that you have is that the San Andreas fault system moves at an average rate of about 3.2 cm/year. Figure 9.7 on the next page shows the interrelationship between recurrence interval, long-term slip rate, and characteristic earthquake magnitude. Knowing any two of these parameters for a given fault system, you can estimate the third. For example, a fault that slips at an average rate of 1 mm/yr and has an earthquake recurrence interval of 1000 years is characterized by earthquakes with magnitudes around 7.0.

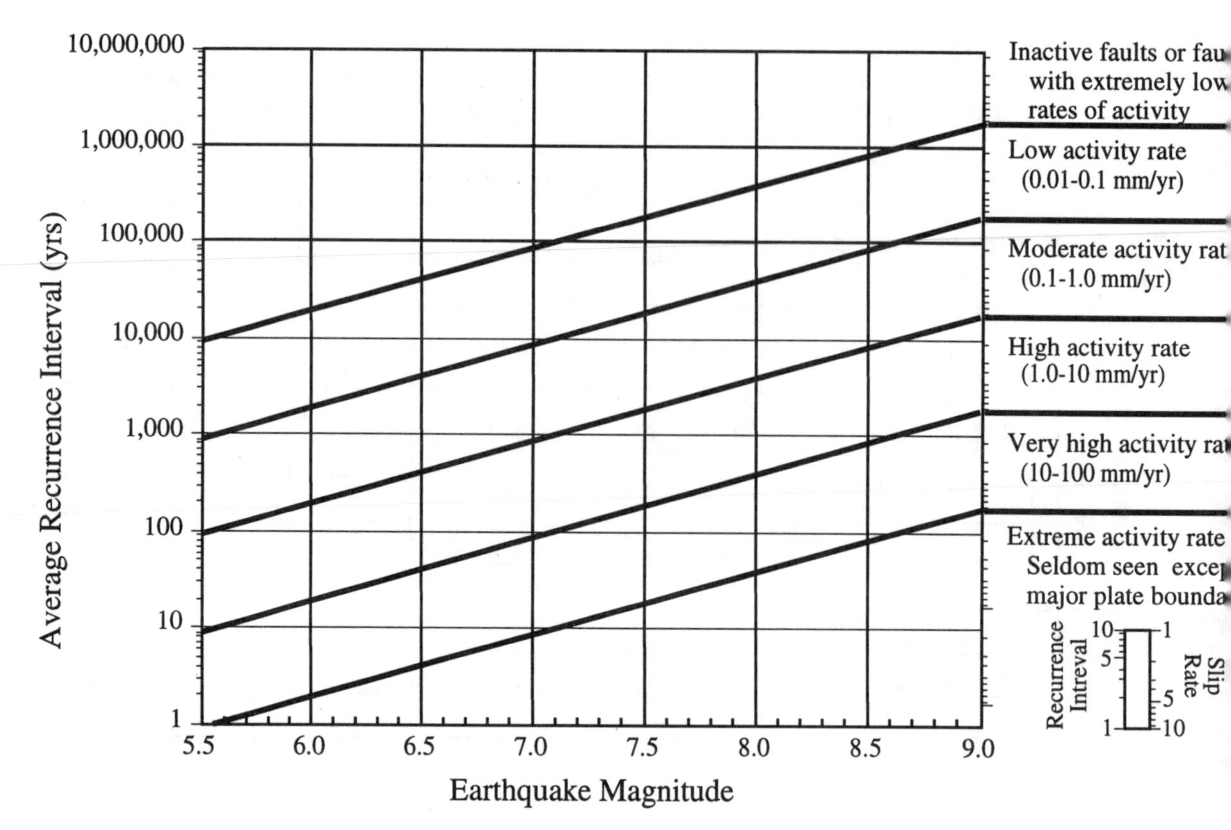

Figure 9.7. General relationship between recurrence interval, long-term slip rate, and characteristic earthquake magnitude. Earthquake magnitude (horizontal axis) is found by plotting the intersection of recurrence interval (vertical axis) and slip rate (diagonal lines). (After Slemmons and DePolo, 1986)

9) Using the long-term slip rate of the San Andreas fault, your estimate of the average recurrence interval at Pallett Creek, and Figure 9.7, what is the characteristic earthquake magnitude for earthquakes along this segment of the fault?

BIBLIOGRAPHY

Lung, R., and R.J. Weick, 1987. Exploratory trenching of the Santa Susana fault in Los Angeles and Ventura Counties. U.S. Geological Survey Professional Paper, 1339: 65-70.

Olig, S.S., W.R. Lund, and B.D. Black, 1994. Large mid-Holocene and late Pleistocene earthquakes on the Oquirrh fault zone, Utah. Geomorphology, 10: 285-315

Sieh, K., 1978. Prehistoric large earthquakes produced by slip on the San Andreas fault at Pallett Creek, California. Journal of Geophysical Research, 83: 3907-3939.

Sieh, K., 1984. Lateral offsets and revised dates of large prehistoric earthquakes at Pallett Creek, Southern California. Journal of Geophysical Research, 89: 7641-7670.

Sieh, K., M. Stuiver, and D. Brillinger, 1989. A more precise chronology of earthquakes produced by the San Andreas fault in southern California. Journal of Geophysical Research, 94: 603-623.

Sims, J.D., 1973. Earthquake-induced structures in sediments of Van Normal Lake, San Fernando, California. Science, 162: 161-163.

Slemmons, D.B., and C. DePolo, 1986. In Active Tectonics. National Academy Press: Washington, D.C.

Acknowledgements: Much of the original data in this exercise and several of the figures come from Sieh (1984) and Sieh (1978). The author would like to thank Dr. Sieh for his permission to use this material.

REGIONAL FOCUS D

THE NEW MADRID SEISMIC ZONE

INTRODUCTION

On December 16, 1811, Scottish naturalist James Bradbury was collecting specimens along the Mississippi River in Missouri when the ground began to shake. Bradbury reported that the ground surface rolled in waves several feet high and that the shaking was accompanied by a sound like "the loudest thunder." This earthquake was the first of three such events – the others occurred on Jan. 23 and Feb. 7 of 1812 – with estimated magnitudes greater than 8.0. These earthquakes caused violent ground cracking and volcano-like eruptions of sediment *(sand blows)* over an area of >10,500 km^2, and uplift of a 50 km by 23 km zone (the Lake County uplift). The shaking rang church bells in Boston, collapsed scaffolding on the Capitol in Washington DC, and was felt over a total area of over 10 million km^2 (the largest felt area of any historical earthquake). Of *all* the historical earthquakes that have struck the U.S., an 1811-style event would do the *most* damage if it recurred today. Most recently, a prediction by an amateur seismologist that a major earthquake would strike the New Madrid seismic zone on or around Dec. 3, 1990 caused a flurry of media attention.

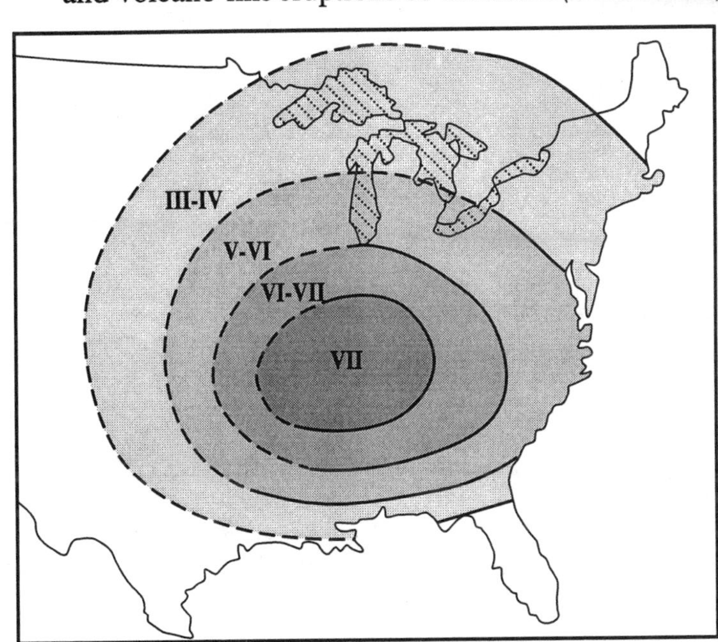

Figure D1. Intensity of shaking during the 12-16-1811 earthquake (Modified Mercalli scale; see Exercise 2). (After Stover and Coffman, 1993)

The New Madrid earthquakes are especially noteworthy because the seismic zone is in the center of the North American Plate. As discussed in Regional Focus B, such *intraplate earthquakes* are felt, and do damage, over much broader areas than comparable earthquakes at plate boundaries. The precise driving force responsible for activity on the New Madrid seismic zone is not known, but most scientists infer that it is compression transmitted across the North American Plate. That compression is focused on New Madrid because it is the site of a Paleozoic structure – the Reelfoot rift – that is a zone of weakness in the crust. The oblique angle between compression and the trend of the rift (Figure D1) apparently causes right-lateral strike-slip motion across the New Madrid seismic zone. Geodetic measurements (precise positioning using GPS satellites) suggest that strain is currently accumulating at about one-third the strain rate of the San Andreas fault!

Strain rates like the one suggested above are a major problem on the New Madrid seismic zone because the structure, the stratigraphy, and the topography of the region show relatively little deformation. Three explanations have been proposed: 1) recent seismological and geodetic activity is still a short-term response to the 1811-12 earthquakes; 2) activity is irregular or cyclic; and 3) activity began only in the recent geologic past. In addition, there is much dispute over how often earthquakes like the 1811-12 sequence occur. Many researchers estimate a recurrence interval of between 550 and 1100 years. On the other hand, other researchers find *no* evidence of great earthquakes in the 5 to 10,000 years prior to 1811. The actual recurrence time of earthquakes like those of 1811-12 has a major effect on the degree of seismic hazard in the central United States.

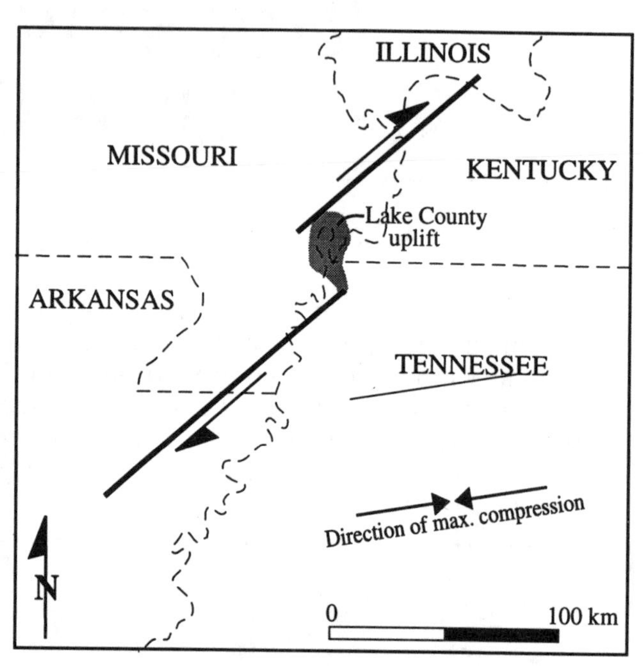

Figure D2. The New Madrid seismic zone. (After Schweig and Ellis, 1994)

GENERAL REFERENCES

Bazell, R., 1990. A little shaky. The New Republic, 203 (Dec. 10): 16-17.

Finkbeiner, A., 1990. California's revenge. Discover, 11 (Sept.): 78-85.

Harris, J.B., and J.D. Kiefer, 1994. Update on the New Madrid seismic zone. Geotimes, 39 (July): 14-18.

Hubbell, S., 1990. Earthquake fever. The New Yorker, 66 (Feb. 11): 75-84.

Kerr, R.A., 1991. A fruitless search for great Midwest quakes. Science, 252: 1497.

Monastersky, R., 1993. Midcontinent heat may explain great quakes. Science News, 143 (May 29): 342.

Stover, C.W., and J.L. Coffman, 1993. Seismicity of the United States, 1568-1989 (Revised): U.S. Geological Survey Professional Paper 1527.

TECHNICAL REFERENCES

Crone, A.J., F.A. McKeown, S.T. Harding, R.M. Hamilton, D.P. Russ, and M.D. Zoback, 1985. Structure of the New Madrid seismic source zone in southeastern Missouri and northeastern Arkansas. Geology, 13: 547-550.

Gomberg, J., and M. Ellis, 1994. Topography and tectonics of the central New Madrid seismic zone: results of numerical experiments using a three- dimensional boundary element program. Journal of Geophysical Research, 99: 20,299-20,310.

Liu, L., M.D. Zoback, and P. Segall, 1992. Rapid intraplate strain accumulation in the New Madrid seismic zone. Science, 257: 1666-1669.

McKeown, F.A., R.M. Hamilton, S.F. Diehl, and E.E. Glick, 1990. Diapiric origin of the Blytheville and Pascola Arches in the Reelfoot Rift, east-central United States: relation to New Madrid seismicity. Geology, 18: 1158-1162.

McKeown, F.A., and L.C. Pakiser, 1982. Investigations of the New Madrid, Missouri, earthquake region. U.S. Geological Survey Professional Paper.

Merritts, D., and T. Hesterberg, 1994. Stream networks and long-term surface uplift in the New Madrid seismic zone. Science, 265: 1081-1084.

Saucier, R.T., 1991. Geoarchaeological evidence of strong prehistoric earthquakes in the New Madrid (Missouri) seismic zone. Geology, 19: 296-298.

Schweig, E.S., and M.A. Ellis, 1994. Reconciling short recurrence intervals with minor deformation in the New Madrid seismic zone. Science, 264: 1308-1311.

Tuttle, M.P., and E.S. Schweig, 1995. Archeological and pedological evidence for large prehistoric earthquakes in the New Madrid seismic zone, central United States. Geology, 23: 253-256.

Zoback, M.D., R.M. Hamilton, A.J. Crone, D.P. Russ, F.A. McKeown, and S.R. Brockman, 1980. Recurrent intraplate tectonism in the New Madrid seismic zone. Science, 209: 971-976.

DISCUSSION QUESTIONS

After reading some of the references listed above, you should be prepared to answer the following questions about the New Madrid seismic zone and the earthquake potential in the central U.S.:

1) The driving mechanism that causes deformation and seismicity in the New Madrid seismic zone is inferred to be compression across the North American plate. Explain how this compression could cause the right-lateral strike-slip faulting illustrated in Figure D2.

2) Explain the geologic setting of the New Madrid seismic zone. Why is earthquake activity concentrated in this location?

3) Each of the New Madrid earthquakes was about the same magnitude as the 1906 San Francisco earthquake. Why, then, do people say that another New Madrid earthquake would do much more damage than another San Francisco earthquake?

4) Outline the evidence that the New Madrid seismic zone is characterized by short recurrence intervals (1100 years or less) for great earthquakes.

5) Outline the evidence that the New Madrid seismic zone is characterized by long recurrence intervals (5-10,000 years or more) for great earthquakes.

EXERCISE 10

CONDITIONAL PROBABILITY

Supplies Needed

- calculator

PURPOSE

Previous exercises in this book have outlined methods for inferring the patterns and history of earthquake activity and faulting. This information is vital for assessing seismic hazard, but in its undigested form, it is not particularly useful to engineers, regional planners, or the general public. Earthquake-hazard is the bridge than connects relatively raw scientific data (fault patterns, slip rates, recurrence intervals, and ages of past earthquakes) with their practical applications. The purpose of this exercise is to illustrate some of the basic principles by which conditional probabilities are calculated.

INTRODUCTION

Conditional probability is defined as the likelihood that a given event – in this case an earthquake – will occur within a specified time period. This likelihood is based on information regarding past earthquakes in a given area and the basic assumption that future seismic activity will follow the pattern of activity observed in the past. Figure 10.1 is an example of a conditional-probability model for Southern California for the period, 1994-2024 (Working Group on California Earthquake Probabilities, 1995). The model gives the percent probability of a large earthquake during this 30-year period on each fault segment shown. Conditional probability is calculated only for those faults for which geologists have collected enough information to make an informed estimate of seismic hazards. This model predicts a 80-90% likelihood that an earthquake with magnitude equal to or greater than 7.0 will strike somewhere in Southern California before 2024, with the single greatest probability coming from the San Jacinto fault, just east of Los Angeles. This information, along with probabilities of maximum seismic shaking in different locations, is used by architects and engineers in designing structures within acceptable safety margins.

Conditional-probability predictions are only as good as the data used to create them. After the Loma Prieta earthquake struck the Santa Cruz mountains, just southeast of San Francisco, in 1989, some geologists called this a success for the conditional-probability

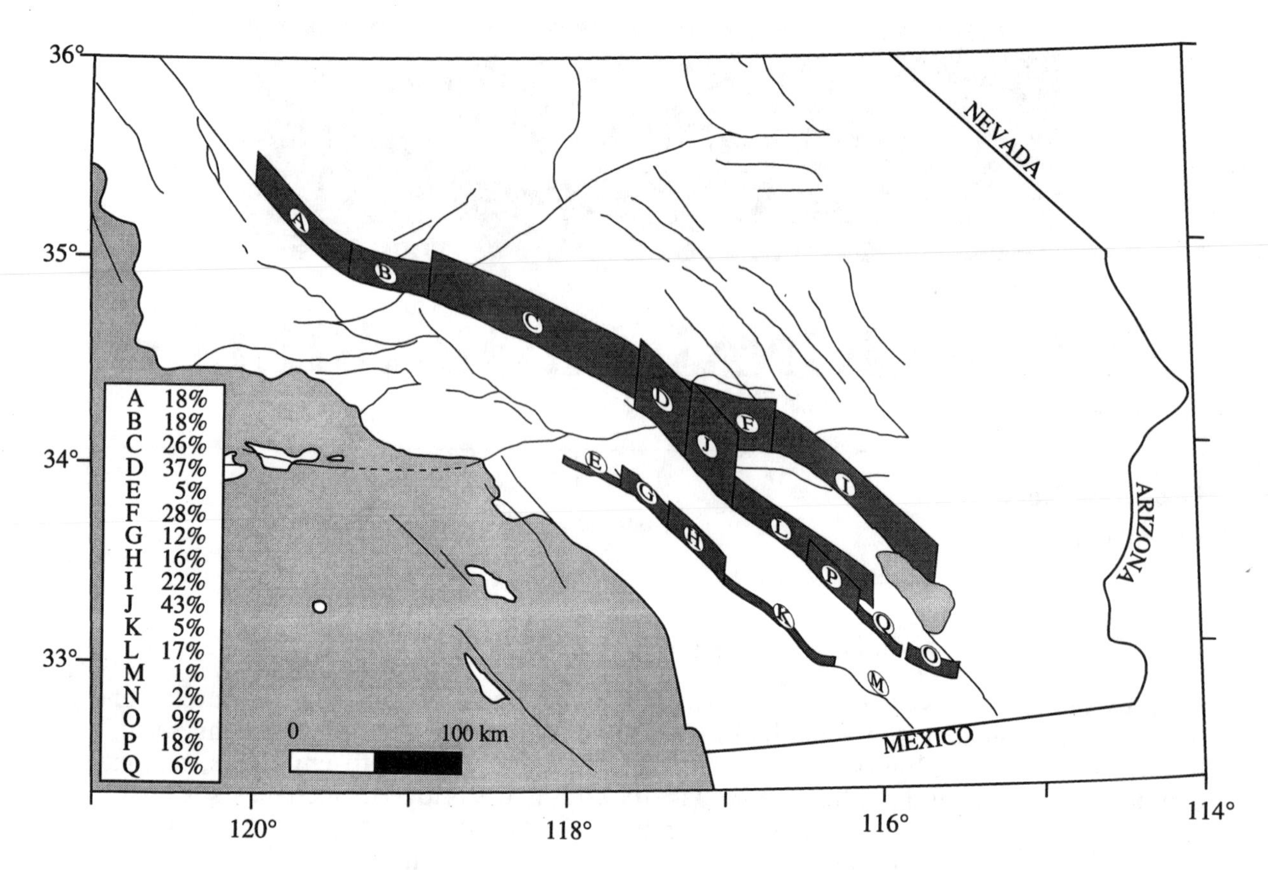

Figure 10.1. Probabilities of fault rupture for the period 1994 to 2024. Width of the shaded bars indicate percent probability for each fault segment. (After Working Group on California Earthquake Probabilities, 1995)

approach. A probability model published in 1988 had assigned this area a 30% chance of a major earthquake in the subsequent 30 years, the second highest value of any segment of the San Andreas fault. At the same time, however, the same model had called this particular probability "equivocal," assigning it the lowest rating on its reliability scale (an E, on a reliability scale from A to E). Many geologists consider such estimates of reliability to be at least as important as the conditional probability itself.

EARTHQUAKE RECURRENCE

Conditional probability is based on models of how and when earthquakes recur. Previous exercises in this book used various types of geologic data to estimate earthquake *recurrence intervals,* the average time between earthquakes on a given fault. In this exercise, we need to examine this concept a bit more closely. Our understanding of earthquake recurrence is fundamentally based on the *elastic-rebound model,* which states that earthquakes occur when elastic strain along a fault exceeds the strength of the rock. Earthquakes release the strain built up during the preceding years.

Figure 10.2 illustrates three different examples of earthquake recurrence, based on three different interpretations of the elastic-rebound model. The first one, the *characteristic-earthquake model,* assumes that a given fault segment is characterized by earthquakes with approximately the same magnitudes and amount of slip. Given a constant long-term strain rate, these characteristic earthquakes would occur at approximately equal intervals. In the elastic-rebound model, characteristic earthquakes would occur only where two strict requirements are met:

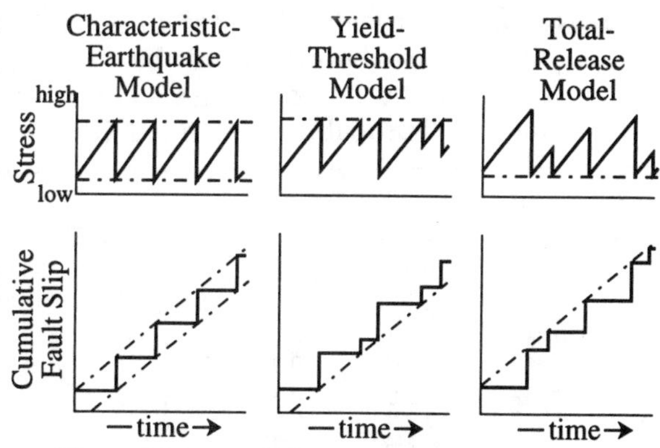

Figure 10.2. Models of earthquake recurrence. (After Shimazaki and Nakata, 1980)

A) the fault has a constant, predictable strain threshold, and earthquakes occur when strain exceeds that threshold

B) earthquakes on the fault release all accumulated strain.

In the *yield-threshold model,* Requirement A is met, but not Requirement B. In the *total-release model,* Requirement B is met, but not Requirement A. In both of these models, earthquakes recur periodically, but with unequal recurrence intervals.

In both the *yield-threshold* and the *total-release models,* recurrence intervals are not constant, but like exam grades in a large university class, the intervals may follow a predictable distribution (Figure 10.3). Various statistical distributions can be used, but this exercise assumes a normal (or *Gaussian)* distribution. Remember that a normal distribution can be described by a mean (μ) and a standard deviation (σ); about 68% of all values fall in the range between $\mu-\sigma$ and $\mu+\sigma$, and 95% fall between $\mu-(2\sigma)$ and $\mu+(2\sigma)$. A statistical probability table (Table 10.1) can be used to find the probability that the next interval between earthquakes will exceed a <u>predicted</u> duration of time (T).

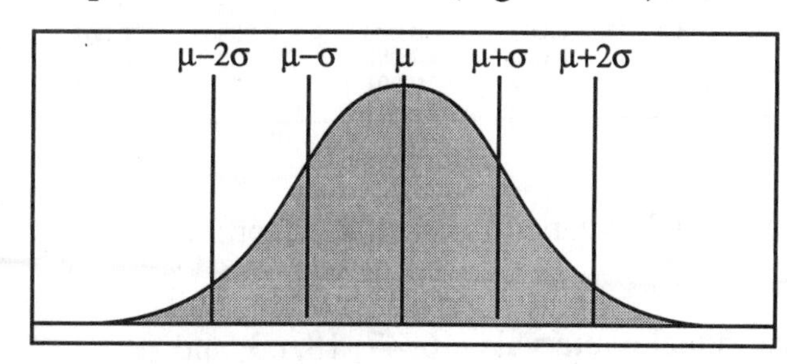

Figure 10.3. A normal distribution.

Table 10.1 is simply a list of 300 probabilities, one for each of 300 *normalized values* (N(T) for normalized time intervals). A normalized value is simply a value that is scaled for the mean and standard deviation of that distribution:

$$N(T) = \frac{T-\mu}{\sigma} \qquad (10.1)$$

For example, $N(\mu+2\sigma) = 2.00$. In order to use Table 10.1, calculate N(T) to two decimal places, and then find the integer and the first decimal (e.g., 2.0 for N(T)=2.00) on the vertical axis of the table and the second decimal (_._0 for N(T)=2.00) on the horizontal axis. Finally, find the value within the table at the intersection of those two axes (0.023 for N(T)=2.00). Remember that this value is the probability from (0.0 to 0.5) that the actual earthquake recurrence interval will exceed T, the predicted interval.

Table 10.1. Probabilities that an actual value will exceed a predicted value, based on a normal distribution. Find the first two digits of N(T) on the vertical axis and the last digit on the horizontal axis. For N(x)<0 (values of T less than μ), *subtract* the indicated probability from 1.000.

N(T)	_._0	_._1	_._2	_._3	_._4	_._5	_._6	_._7	_._8	_._9
0.0_	0.500	0.496	0.492	0.488	0.484	0.480	0.476	0.472	0.468	0.464
0.1_	0.460	0.456	0.452	0.448	0.444	0.440	0.436	0.433	0.429	0.425
0.2_	0.421	0.417	0.413	0.409	0.405	0.401	0.397	0.394	0.390	0.386
0.3_	0.382	0.378	0.375	0.371	0.367	0.363	0.359	0.356	0.352	0.348
0.4_	0.345	0.341	0.337	0.334	0.330	0.326	0.323	0.319	0.316	0.312
0.5_	0.309	0.305	0.302	0.298	0.295	0.291	0.288	0.284	0.281	0.278
0.6_	0.274	0.271	0.268	0.264	0.261	0.258	0.255	0.251	0.248	0.245
0.7_	0.242	0.239	0.236	0.233	0.230	0.227	0.224	0.221	0.218	0.215
0.8_	0.212	0.209	0.206	0.203	0.201	0.198	0.195	0.192	0.189	0.187
0.9_	0.184	0.181	0.179	0.176	0.174	0.171	0.169	0.166	0.164	0.161
1.0_	0.159	0.156	0.154	0.152	0.149	0.147	0.145	0.142	0.140	0.138
1.1_	0.136	0.134	0.131	0.129	0.127	0.125	0.123	0.121	0.119	0.117
1.2_	0.115	0.113	0.111	0.109	0.108	0.106	0.104	0.102	0.100	0.099
1.3_	0.097	0.095	0.093	0.092	0.090	0.089	0.087	0.085	0.084	0.082
1.4_	0.081	0.079	0.078	0.076	0.075	0.074	0.072	0.071	0.069	0.068
1.5_	0.067	0.066	0.064	0.063	0.062	0.061	0.059	0.058	0.057	0.056
1.6_	0.055	0.054	0.053	0.052	0.051	0.050	0.049	0.048	0.047	0.046
1.7_	0.045	0.044	0.043	0.042	0.041	0.040	0.039	0.038	0.038	0.037
1.8_	0.036	0.035	0.034	0.034	0.033	0.032	0.031	0.031	0.030	0.029
1.9_	0.029	0.028	0.027	0.027	0.026	0.026	0.025	0.024	0.024	0.023
2.0_	0.023	0.022	0.022	0.021	0.021	0.020	0.020	0.019	0.019	0.018
2.1_	0.018	0.017	0.017	0.017	0.016	0.016	0.015	0.015	0.015	0.014
2.2_	0.014	0.014	0.013	0.013	0.013	0.012	0.012	0.012	0.011	0.011
2.3_	0.011	0.010	0.010	0.010	0.010	0.009	0.009	0.009	0.009	0.008
2.4_	0.008	0.008	0.008	0.008	0.007	0.007	0.007	0.007	0.007	0.006
2.5_	0.006	0.006	0.006	0.006	0.006	0.005	0.005	0.005	0.005	0.005
2.6_	0.005	0.005	0.004	0.004	0.004	0.004	0.004	0.004	0.004	0.004
2.7_	0.003	0.003	0.003	0.003	0.003	0.003	0.003	0.003	0.003	0.003
2.8_	0.003	0.002	0.002	0.002	0.002	0.002	0.002	0.002	0.002	0.002
2.9_	0.002	0.002	0.002	0.002	0.002	0.002	0.002	0.001	0.001	0.001

Example 10.1:

Find the probability that a fault will rupture in the next 25 years if the fault has ruptured in 1729, 1775, 1822, 1840, 1903, and 1969.

These six earthquakes define five inter-earthquake intervals, 46, 47, 18, 63, and 66 years long. The mean of these intervals is:

$$\mu = \frac{46 + 47 + 18 + 63 + 66}{5} = 48 \text{ years}$$

The standard deviation of these intervals is:

$$\sigma = \frac{|46-48| + |47-48| + |18-48| + |63-48| + |66-48|}{5}$$

$$\sigma = \frac{2 + 1 + 30 + 15 + 18}{5} = 13.2 \text{ years}$$

Thus the recurrence-interval distribution for this fault is 48±13.2 years. If you are doing this exercise in 1996, you know that the current recurrence interval (since 1969) is at least

27 years. The question here is: what is the likelihood that the fault will rupture in the next 25 years (between 1996 and 2021)? The probability that this recurrence interval will fall between 27 and 52 years is:

$$P[27\text{-}52] = P(27) - P(52), \tag{10.2}$$

which simply says that the probability of the recurrence interval falling in the range, 27 to 52 years, equals the probability of the interval exceeding 27 years minus the probability that it will exceed 52 years. You find P(27) and P(52) by finding the normalized values (Equation 10.1) and then using Table 10.1:

$$N(52) = \frac{T - \mu}{\sigma} = \frac{52 - 48}{13.2} = 0.30$$

$$N(27) = \frac{T - \mu}{\sigma} = \frac{27 - 48}{13.2} = -1.59$$

The probability – P(52) – on Table 10.1 that corresponds to N(T)=0.30 is 0.382. Finding P(27) is only slightly more complicated, because N(27) is negative. As Table 10.1 instructs you, find the value in the table that corresponds to +N(x), and then subtract that value from 1.000:

$$\text{for } N(T) = -1.59,\ P(T) = 1.000 - 0.056 = 0.944$$

Using the values of P(27) and P(52) and Equation 10.2:

$$P[27\text{-}52] = P(27) - P(52) = 0.944 - 0.382 = 0.562$$

This means that there is a 56.2% likelihood that a major earthquake will occur on this fault in the 25-year period between 1996 and 2021.

1) Ground-rupturing earthquakes have occurred on the Parkfield segment of the San Andreas fault in 1857, 1881, 1901, 1922, 1934, and 1966. What is the probability that another earthquake will **not** have occurred at Parkfield between 1966 and the present date?

Conditional Probability

The Parkfield segment of the San Andreas fault is noteworthy because it has ruptured so regularly in the past that, in 1985, the U.S. Geological Survey made a formal prediction that there was a 90% probability that the Parkfield segment would generate a magnitude 5.5-6.0 earthquake by 1993. The fault has remained embarrassingly quiet ever since. In fact, this author hesitates to even mention Parkfield for fear that the fault segment will rupture the day after this book goes to press. The Earth is complex, and there are a number of reasons why a fault might deviate from the recurrence predicted by the past earthquakes, including:

- Earthquake clustering – Some faults are characterized by periods of closely-spaced earthquakes, followed by periods of inactivity.
- Fault-segment triggering – Different segments of the same fault are not truly independent. An earthquake on one segment often triggers rupture on adjacent segments.
- Inadequate period of record – All earthquake-recurrence models are studies in small numbers. Statisticians prefer distributions that consist of hundreds of values, but seismologists rarely have data for more than a handful of past recurrence intervals.

THE WASATCH FAULT ZONE

The 343 km-long Wasatch fault zone marks the eastern boundary of the Basin and Range province, separating the region from the Colorado Plateau and the Middle Rocky Mountains to the east. The Wasatch fault zone underlies a populated corridor that is home to 80% of the inhabitants of Utah. The fault zone is subdivided into ten distinct segments (Figure 10.4), including six more-active central segments and four less-active segments on the margin of the fault zone. All of the segments except the Brigham City segment have ruptured in the last 1500 years, but none in historical time.

Extensive trenching has been done at sites along the Wasatch fault zone to characterize the history of earthquakes in the recent geologic past. Exercise 9 in this book outlines how fault trenches are used in studies of paleoseismology, and Figure 10.5 illustrates one trench cut across the Wasatch fault. The right-hand side of Figure 10.4 summarizes the available information on the number and timing of earthquakes on the six central fault segments during the past 6000 years. In the questions on the following pages, you will use the fault-trench information (the best published data currently available) to estimate conditional probabilities for this fault zone.

2) Summarizing the age information in Figure 10.4, trenches along the Wasatch fault zone record earthquake recurrence intervals of:

   1200, 525, 1725, 1050, 3975, 2100, 2625, 2250, 3675, and 900 years

   Find the mean and standard deviation of this recurrence-interval population.

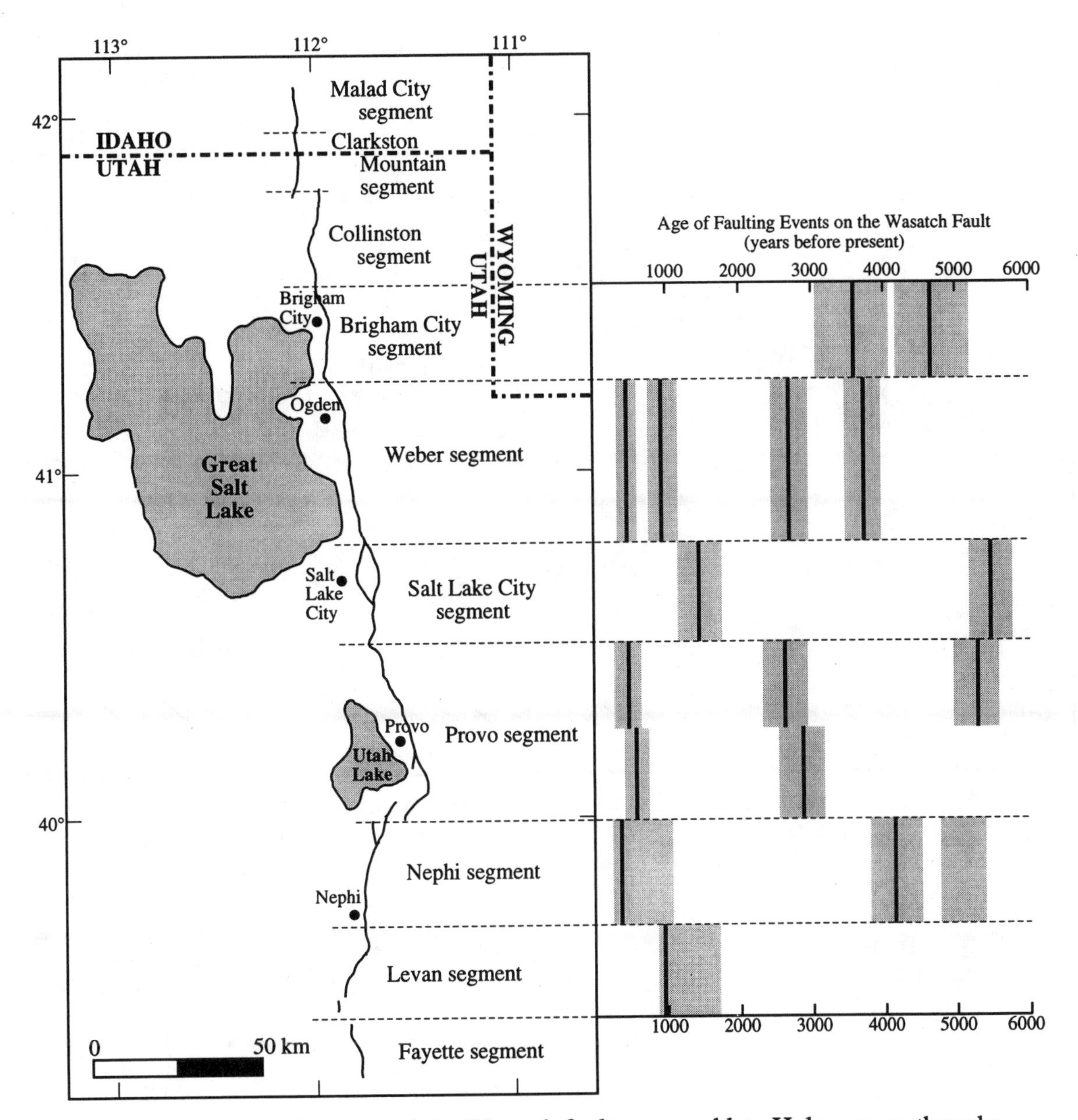

Figure 10.4. Location map of the Wasatch fault zone and late Holocene earthquake history from fault trenches. (After Gori and Hays, 1991)

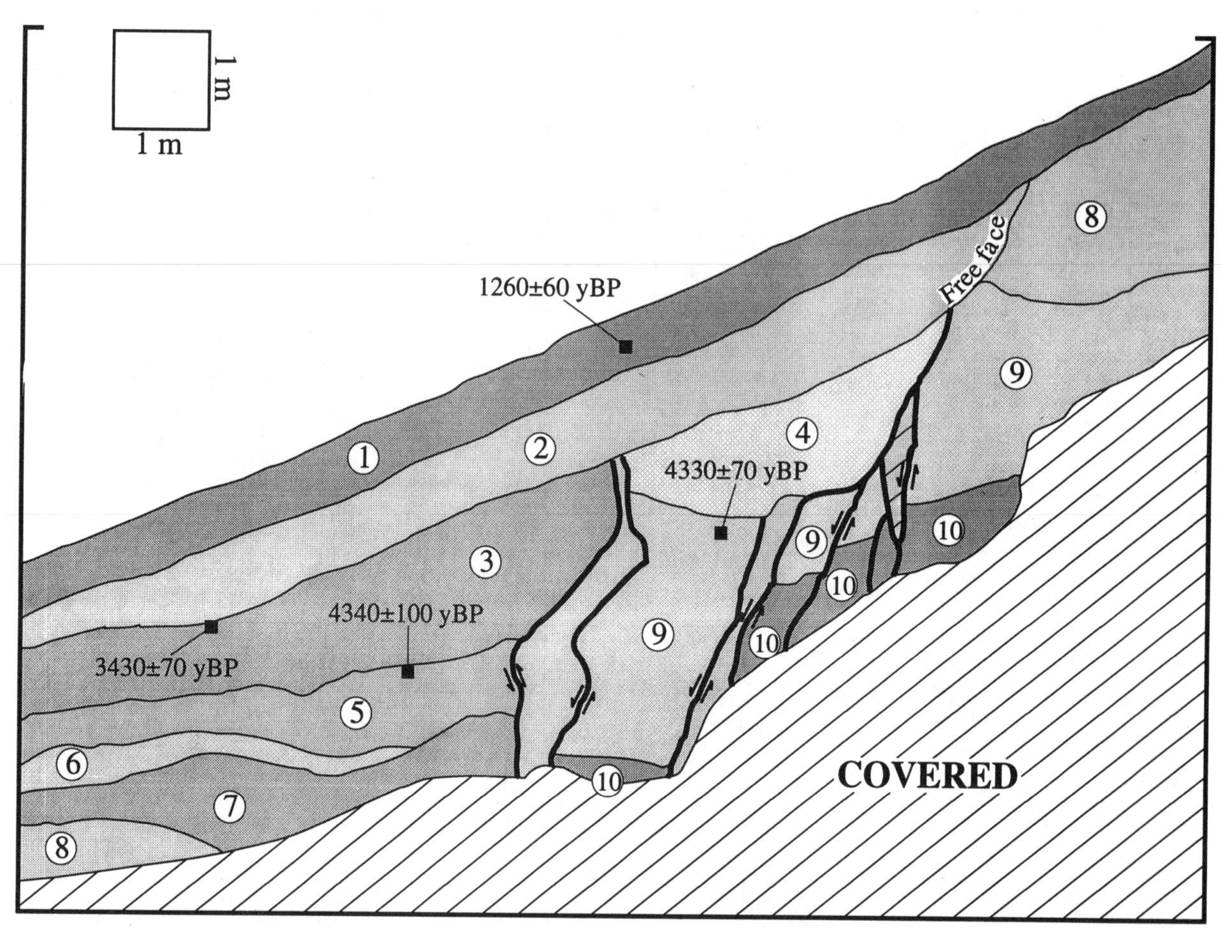

Figure 10.5. Part of trench BC-1 on the Brigham City segment of the Wasatch fault zone illustrating an earthquake that ruptured the ground surface just after 3430 years BP. (After Gori and Hays, 1991)

3) All of the Wasatch fault segments have ruptured in the last 1500 years except the Brigham City segment. The last ground-rupturing earthquake there occurred approximately 3500 years ago. Find the probability that a major earthquake will occur on the Brigham City segment during the next 25 years.

4) The Salt Lake City segment of the Wasatch fault zone last ruptured about 1500 years ago. Find the probability that a major earthquake will occur on this segment during the next 25 years.

5) Explain why the conditional probability for the Brigham City segment is so different from that of the Salt Lake City probability.

CONDITIONAL PROBABILITY AND GROUND SHAKING

Recurrence-interval information helps define the probability of future earthquakes, but it says nothing about their magnitudes of their effects. That information must come from other sources, such as fault slip rates, past dimensions of rupture, past earthquake magnitudes, and site-specific conditions. Combined with recurrence probabilities, this additional information allows us to calculate probabilities of different levels of seismic ground shaking. Figure 10.6 is a regional map of ground-shaking hazard across the U.S., based on recurrence-interval and ground-shaking probabilities. Estimates of ground-shaking risk are vital to architects, engineers, and planners in earthquake-prone areas.

Like recurrence intervals, ground-shaking estimates for a specific site (usually expressed as acceleration) can be described by a mean value and associated standard deviation (a normal distribution). Such estimates must be firmly based on shaking intensities during past earthquakes or other data. The probability of seismic shaking exceeding a given value of acceleration (A) in a pre-specified duration of time (T) is expressed as:

$$P(A,T) = P(A) * P(T). \qquad (10.3)$$

Conditional Probability

P(A,T) is the probability that *both:* 1) an earthquake will occur in the pre-specified time, and 2) the ground shaking will exceed an acceleration of A during that earthquake.

Example 10.2:

A site is characterized by seismic shaking of 0.5±0.3 g (50%±30% of the acceleration of gravity) as a result of rupture on a nearby fault. If there is a 12% chance of the fault rupturing in the next 50 years, what is the probability that this site will experience seismic acceleration greater than 0.7 g in that 50-year period?

The first step in this problem is to find P(A), which is the probability of A (0.7 g) being exceeded during any one earthquake. In order to do this, first calculate N(A):

$$N(A) = \frac{0.7\ g - \mu}{\sigma} = \frac{0.7\ g - 0.5\ g}{0.3\ g} = 0.67$$

using Table 10.1, P(A) = 0.251 = 25.1%

Now combine this information with the probability of recurrence (P(50)=12%):

$$P(A,T) = P(A) * P(T) = 0.251 * 0.120 = 0.030$$

This means that there is a 3.0% likelihood of an earthquake occurring on this fault and causing ground acceleration greater than 0.7 g at the site in question.

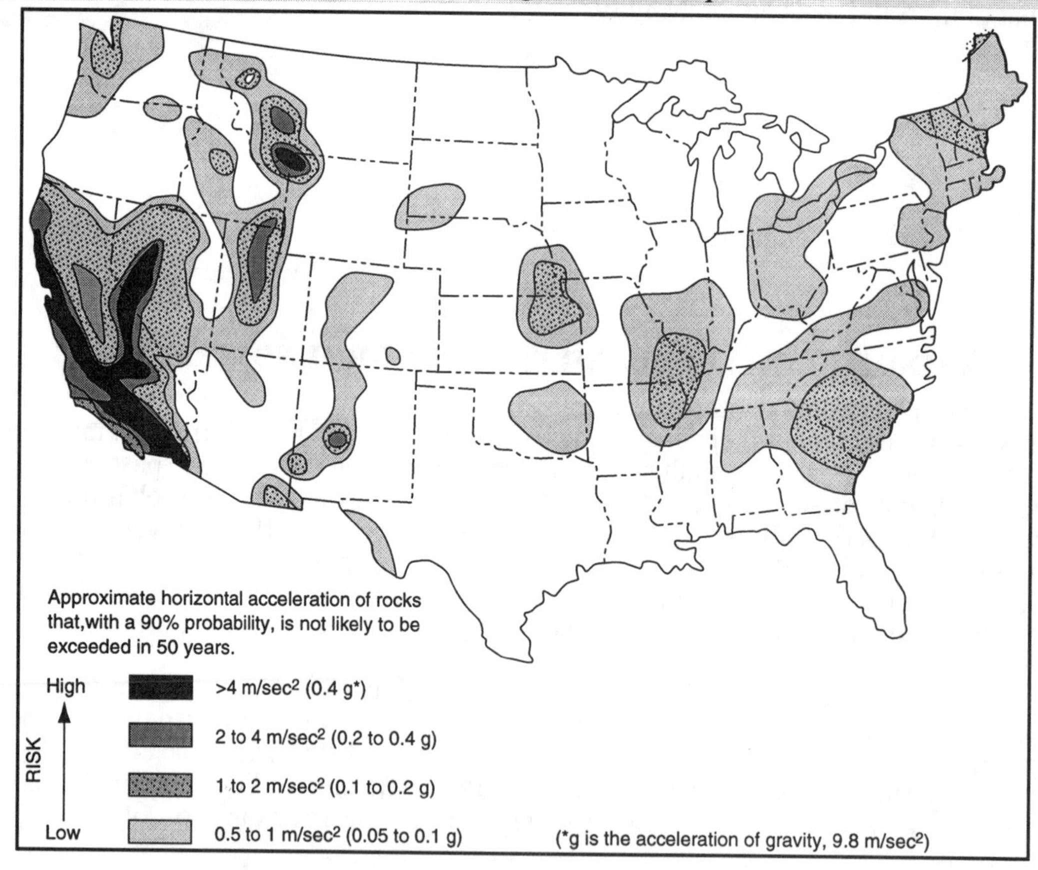

Figure 10.6. Ground-shaking hazards in the U.S. (From Algermissen and Perkins, 1976)

6) A new construction project is being planned for a site 10 km from the Salt Lake City segment of the Wasatch fault zone. That site is characterized by seismic ground acceleration of 0.4±0.2 g (for a M=7.0 earthquake; Joyner and Boore, 1988). Find the probability that the site will experience accelerations greater than 0.7 g during the next 100 years.

7) If the same site in Question 6 were instead 10 km from the Brigham City segment, what would be the probability of exceeding 0.7 g in a 100-year period?

BIBLIOGRAPHY

Algermissen, S.T., and D.M. Perkins, 1976. A probabilistic estimate of maximum acceleration in rock in the contiguous United States. U.S. Geological Survey Open-File Report, 76-416.

Gori, P.L., and W.W. Hays (eds.), 1992. Assessment of regional earthquake hazards and risk along the Wasatch front, Utah. U.S. Geological Survey Professional Paper, 1500-A-J.

Jacob, K.H., 1984. Estimates of long-term probabilities for future great earthquakes in the Aleutians. Geophysical Research Letters, 11: 295-298.

Joyner, W.B., and D.M. Boore, 1988. Measurement, characterization, and prediction of strong ground motion. Proceedings of Earthquake Engineering and Soil Dynamics conference. American Society of Civil Engineers, 43: 102.

Machette, M.N., S.F. Personius, and A.R. Nelson, 1992. The Wasatch Fault Zone, USA. Annales Tectonicae, 6 (Suppl.): 5-39.

Machette, M.N., S.F. Personius, A.R. Nelson, D.P. Schwartz, and W.R. Lund, 1991. The Wasatch fault zone, Utah - segmentation and history of Holocene earthquakes. Journal of Structural Geology, 13: 137-149.

Savage, J.C., 1992. The uncertainty in earthquake conditional probabilities. Geophysical Research Letters, 19: 709-712.

Schwartz, D.P., and K.J. Coppersmith, 1984. Fault behavior and characteristic earthquakes – Examples from the Wasatch and San Andreas fault zones. Journal of Geophysical Research, 89: 5681-5698.

Shimazaki, K., and T. Nakata, 1980. Time-predictable recurrence model for large earthquakes. Geophysical Research Letters, 7: 279-282.

Working Group on California Earthquake Probabilities, 1995. Seismic hazards in Southern California: Probable earthquakes, 1994 to 2024. Bulletin of the Seismological Society of America, 85: 379-439.

Acknowledgements: This exercise was modified from, but largely based on, a conditional-probability exercise created by Ronald Bruhn of the University of Utah. The author gratefully thanks Dr. Bruhn for his permission to use this material.

EXERCISE 11

SEISMIC SHAKING AND EARTHQUAKE ENGINEERING

Supplies Needed

- calculator
- metric ruler
- colored pencils (at least three colors)

PURPOSE

The purpose of the exercise is to familiarize you with the effects of earthquakes on Earth materials and on buildings. The strength of shaking during an earthquake depends on the amplitude of seismic waves that reach the site, the earth materials at and near the surface, and the design of buildings and other structures. The purpose of this exercise is to outline the methods for describing and measuring seismic shaking, and for predicting its effects on human structures.

SEISMIC SHAKING

It's simple to measure the magnitude of an earthquake, but the pattern of shaking that results from a single earthquake (called *seismic shaking,* or *seismic ground motion)* is much more complex. At the same time, predicting ground motion is crucial for designing structures that will withstand the earthquakes likely to occur in an area. Several factors have to be taken into account to be able to predict ground shaking:

- tectonic framework:
  - number of faults in area, and their distance to the site
  - types of faults
  - earthquake recurrence intervals
  - predicted earthquake magnitudes
- near-surface geology
  - type of material
  - thickness of unconsolidated material
- construction materials and techniques.

Seismic Shaking and Earthquake Engineering

Earthquake intensity, introduced in Exercise 2, is an after-the-fact assessment of the strength and pattern of seismic shaking. The magnitude of earthquakes before the existence of seismometers can only be estimated indirectly. Earthquake intensities, however, can be estimated from historical records. The following relationship between magnitude and maximum intensity has been proposed (Howell, 1973) for shallow-focus earthquakes:

$$I_{max.} = (2 * M) - 4.6 \tag{11.1}$$

where I_{max} is the maximum intensity, and M is the magnitude of the earthquake.

One problem with intensity measurements is that they are essentially qualitative, whereas engineers and architects require more quantitative values. Ground motion is most commonly calculated as *acceleration*, expressed as a fraction of the acceleration of gravity (g, where g=9.8 m/sec^2). In addition, seismic acceleration can be broken down into its three perpendicular components: a vertical component (up-down shaking), and two horizontal components (typically east-west shaking, and north-south shaking). The horizontal components of shaking commonly are the most damaging to buildings because structures are already designed to withstand the vertical force of gravity.

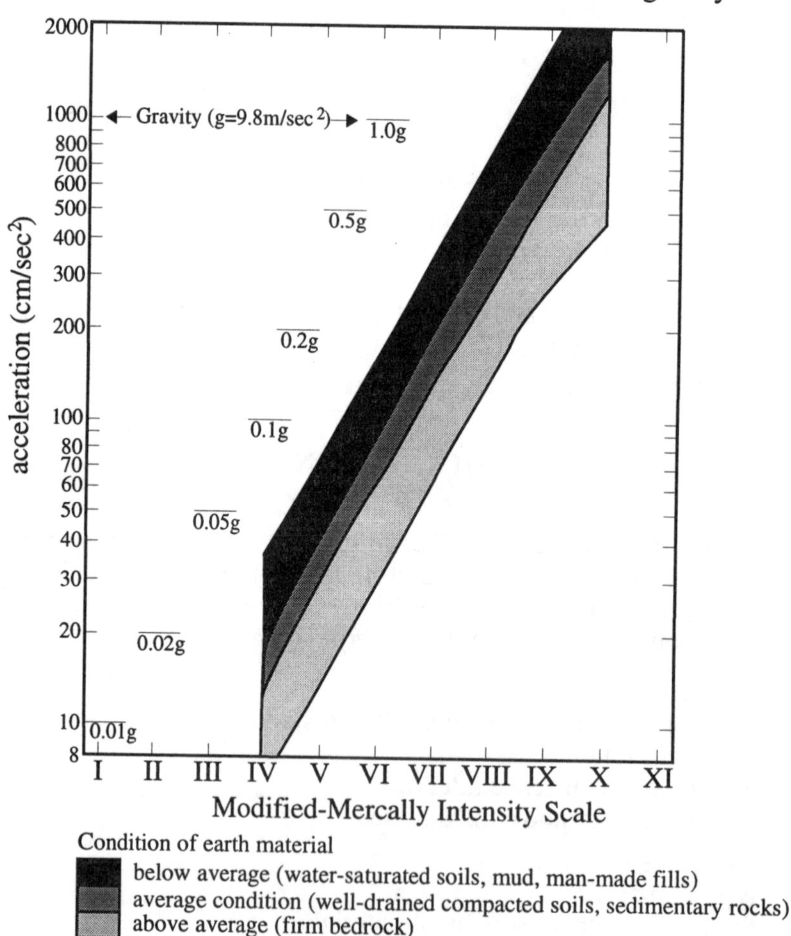

Figure 11.1. Seismic acceleration vs. ground shaking intensity for different earth materials. (After Leed, 1973)

2) Using Figure 11.1 and Equation 11.1, find the peak ground acceleration (in g) on bedrock in San Francisco during an magnitude 7.0 earthquake.

3) How much greater is the acceleration for man-made fills during a magnitude 7.0 earthquake in this area?

Acceleration at a particular site is related to the distance from the site to the epicenter. In addition, acceleration is related to earthquake magnitude – the greater the earthquake, the bigger the acceleration at all locations. An empirical equation (i.e., based on observation and measurement) relates peak ground acceleration (A, in m/sec²) to earthquake Magnitude (M) and distance to the hypocenter (R, in km):

$$A = 1080\, e^{0.5\,M} / (R+25)^{1.32} \qquad (11.2)$$

Remember that e is the natural exponent function (see Exercise 4 for examples of the function). This equation was created for California, although it may be applicable in other plate-boundary settings. As discussed in Regional Focus B, however, intraplate earthquakes are transmitted farther.

MATERIAL AMPLIFICATION

After tectonic framework, the second major factor that determines the seismic hazard at a specific location is the type of material under foot. On a local scale, the intensity of seismic ground motion is mainly a function of surficial geology. The earthquake in 1985 that killed 5600 people in Mexico City was actually centered several hundred kilometers away in the Pacific Ocean. Many coastal areas much closer to the epicenter than Mexico City, including the city of Acapulco, sustained far less damage. Destruction in Mexico City was so severe because the city is built upon a thick pile of loose lake sediments which *amplify* seismic shaking.

Material amplification is defined as increased seismic shaking as a result of surficial and near-surface deposits. Amplification is a function of the composition and thickness of loose material that underlies an area In particular, sediments and thick soils tend to amplify seismic waves.

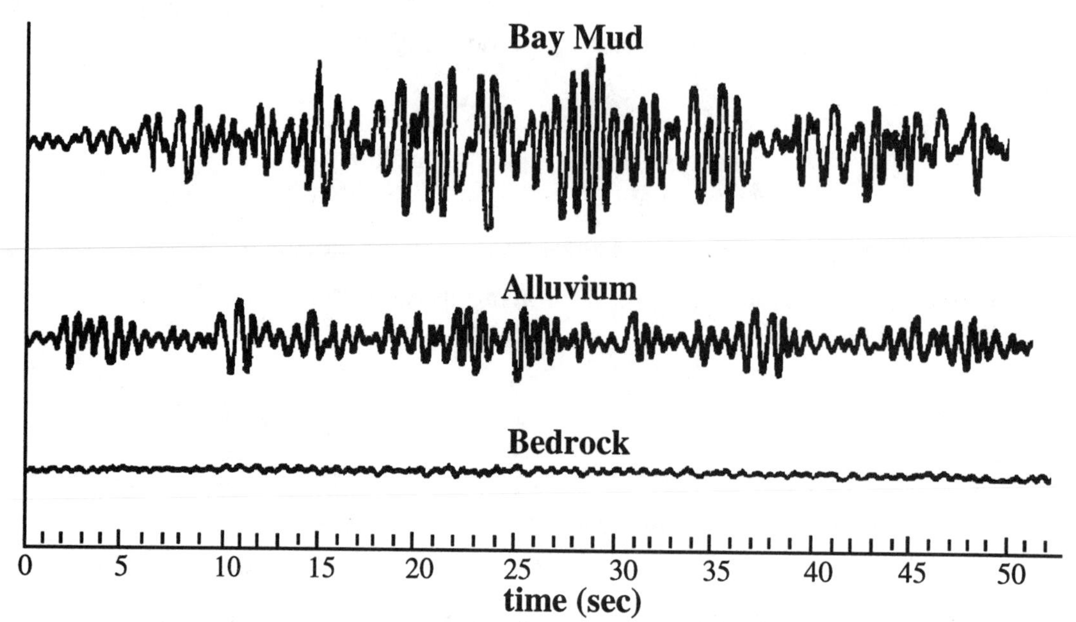

Figure 11.2. Horizontal ground motions of an underground nuclear explosion were recorded by accelographs in San Francisco. All materials were subjected to the same seismic waves. (After Borcherdt, 1975)

4) The vertical axis in Figure 11.2 uses the same scale for all three plots. Calculate how many times more violent the ground shaking was on A) bay mud and B) alluvium compared to bedrock.

A)

B)

In the terminology of seismic waves, material amplification increases the *amplitude* of ground motion, but the *period* of these waves is also important. Like the amplitude, the period of ground motion is a function of both the seismic source and the local geologic conditions. Different types of Earth materials each tend to vibrate at a characteristic period, known as the *fundamental period* of that material. Buildings also have a fundamental period, and the fundamental periods of different buildings vary depending on the height of the structure and other factors (Figure 11.4).

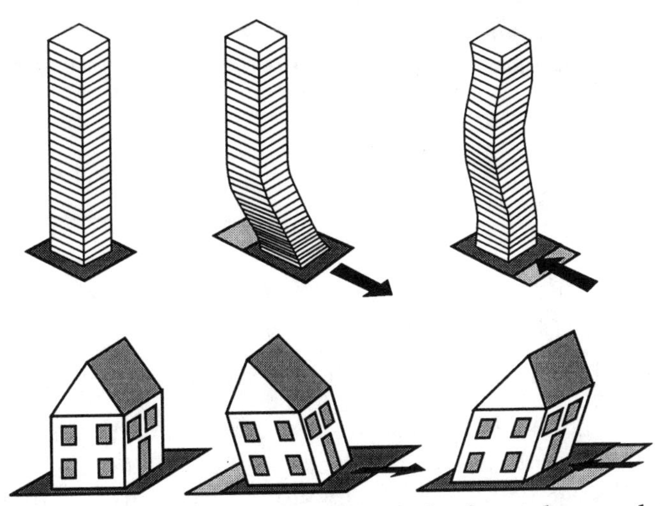

Figure 11.3. The response of a tall building to horizontal ground shaking compared with the response of a short building. (After Lagorio, 1990)

An important factor in predicting earthquake damage is the relationship between the fundamental period of a building and the period of the material on which the building is constructed. If the building's period equals the fundamental period of the material on which it is built, or if it equals some whole-number multiple of the material's fundamental period, then seismic shaking will create a *resonance* with the building that can greatly increase the stresses on the structure. Tall buildings tend to be damaged more on deep, soft soils because of their similar vibrational period. Small, rigid buildings perform poorly on short-period materials such as bedrock (Figure 11.4).

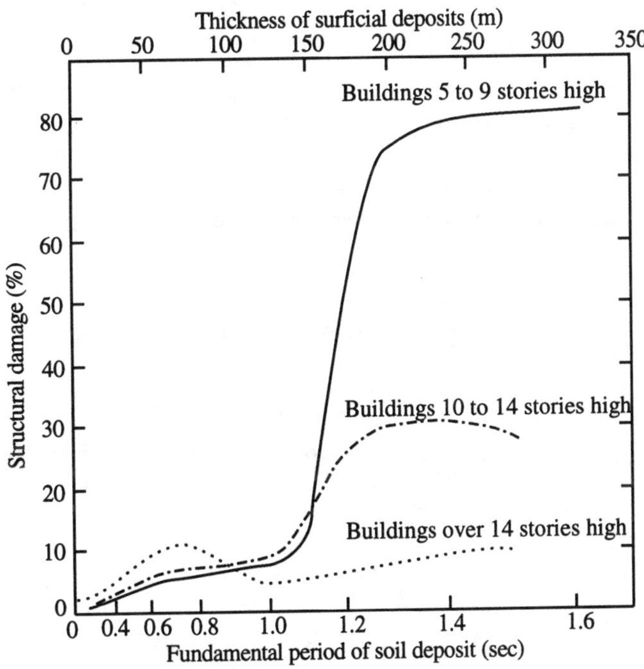

Figure 11.4. Relationship between building height, thickness of sediments, and earthquake damage. (After Seed, 1972)

5) At what fundamental period (in seconds) is the damage greatest for buildings of:

a) 5-9 stories?

b) 10-14 stories?

c) over 14 stories?

SECONDARY EFFECTS OF EARTHQUAKES

Seismic shaking is considered the *primary effect* of earthquakes, but the majority of damage and casualties often result from the indirect, *secondary effects*. The earthquake that struck San Francisco is 1906 is often called the "Great Fire" of 1906 because it was the three days of fire that followed the earthquake that did the great majority of damage, not the seismic shaking itself. Major secondary effects of earthquakes include:

- fire
- landslides
- tsunami
- liquefaction

All of the effects listed above can cause major damage during or shortly after an earthquake. The best guide to where this type of damage will occur in the future is where it has occurred in the past. Landslide-hazard, tsunami-hazard, and liquefaction-hazard maps have been compiled in many of the areas that face the greatest likelihood of these threats in the future.

We will focus on one of these secondary hazards because it is closely linked to seismic ground motion. *Liquefaction* occurs in fine, water-saturated sands during strong seismic shaking. Water-saturated sediments lose their strength as the grains are rearranged during shaking and fluid pressure increases. A sixteenth-century earthquake that struck Port Royal, Jamaica caused much of the town and its inhabitants simply to dissappear into the ground.

BUILDING CODES

In designing earthquake-safe structures, the most important criteria is to make sure that no one critical element of the building is overstressed (the "weakest link" philosophy). In other words, the building is only as strong as its weakest link. The standard in designing a safe building is the capacity or strength of the material to resist seismic stresses, in particular horizontal acceleration, without failure. It's important to understand that seismic provisions in most building codes are intended to protect life and reduce property damage but not completely eliminate losses. The Structural Engineers Association of California Structures recommend that structures be able to (a) resist minor earthquakes without damage, and (b) resist moderate earthquakes without structural damage but with some non-structural damage.

Buildings that are most vulnerable to lateral forces induced by seismic waves are unreinforced masonry, brick and mortar, and adobe constructions. Small wood frame structures are usually the safest as long as they are securely anchored to their foundations. Houses that are not anchored or are improperly anchored can shear off their foundations during lateral ground motion. Steel frame or reinforced-concrete construction methods are least hazardous for multi-story buildings or other tall structures.

The horizontal components of seismic shaking can be converted into a parameter called *base shear.* As mentioned earlier, it is the horizontal components of acceleration that are potentially the most damaging to buildings. Base shear is the maximum lateral force imposed on a structure during seismic ground motion. Newton's second law states that force and acceleration are related as follows:

$$F = m * a \tag{11.3}$$

where F is the force in Newtons (1 Newton = 1 N = 1 kg*m/sec^2), m is the mass (in kg) and a is acceleration in g (1 g = the acceleration of gravity = 9.8 m/sec^2). In this case, F is the base shear, m is the mass of the structure, and a is the maximum horizontal component of seismic acceleration.

The actual base shear that a building experiences during an earthquake can be calculated using Equation 11.3. A much more applied equation is used by architects and engineers to calculate the base shear that a structure at a specific site *should* withstand:

$$BS = Z * I * (C/R_w) * m * g \tag{11.4}$$

where BS is the base shear, Z is the seismic zone factor (a unitless value that incorporates the maximum seismic shaking at different locations), I is the importance factor for a structure (high for a school, for example, and low for a warehouse), C is a numerical coefficient related to soil conditions, R_w is a parameter that assessed the type of construction used, m is the total mass of the structure (in kg), and g is gravity (9.8m/sec^2) (Lagorio, 1990).

A building resists base shear with the strength of its load-bearing walls *(shear walls).* During east-west-oriented shaking, or when the east-west component of shaking is the greatest component, the north-south walls of building must withstand the majority of the stress. The number and arrangement of shear walls determines the portion of base shear that each wall must withstand. Solid, regular arrangements of walls tend to be the strongest; buildings with unsupported walls, such as ground-level carports, often are the most vulnerable.

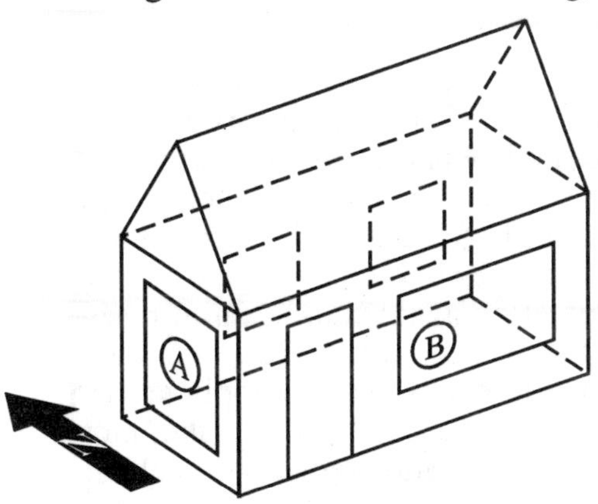

Figure 11.5. Shear walls resist base shear, supporting a building during horizontal shaking.

The building in Figure 11.5 consists of two pairs of perpendicular walls. Walls A and B each must be designed to withstand <u>one half</u> of the total base shear. If shaking is predominantly east-west, then Wall B is one of two walls resisting the seismic stress; if shaking is predominantly north-south, then Wall A is in the same situation.

6) Suggest simple modifications or additions to the building in Figure 11.5 so that both Wall A and Wall B would only need to withstand one third of the base shear.

CASE STUDY: EARTHQUAKES IN SAN FRANCISCO

The San Francisco Peninsula lies astride the San Andreas fault where it passes northward from the land to the ocean. In 1906, the city was burgeoning as the preeminent cultural and economic center west of the Mississippi. At 5:14 a.m. on April 18, the city's fortunes suffered a severe setback:

> *"At first came a sharp but gentle swaying motion that grew less and less; then a heavy jolting sideways – then another, heaviest of all. Finally a grinding round of everything, irregularly tumultuous, spasmodic, jerky."* (Aitken and Hilton, 1906)

Damage was greatest in the Marina District of the city, where liquefaction and settling in the underlying soft sediments and artificial fill crumbled many buildings. In most of the city, however, damage was minor, and some residents simply returned to bed. Over 70% of the damage done in 1906 occurred in the fire that followed the earthquake. In the Marina district and other pockets of damage, stoves and lamps toppled, igniting scattered blazes. Firefighters rushed to contain the fires, but watched water pressure drop to a trickle and then to nothing because water mains had been cut during the shaking. For four days, fire ravaged the city, consuming 490 city blocks.

The 1906 earthquake and fire were not the first such events to strike San Francisco. The city was extensively damaged during an earthquake in 1868, in which it was again the Marina District that suffered the most. In addition, fire had swept through the city three times previously: in 1849, 1850, and 1851. The extent to which San Francisco had learned its lessons were again tested in 1989. The Loma Prieta earthquake, a magnitude 7.1 shock centered about 120 km southeast of the city, caused shaking damage similar in pattern to the damage in both 1868 and 1906. Liquefaction again caused buildings to collapse in San Francisco's Marina District, igniting scattered fires, and broke a number of water mains. Fortunately, technical improvements in the city's water-supply system avoided a repeat of the 1906 fire. Of the 65 fatalities in the 1989 earthquake, most occurred when a mile-long (1.6 km) segment of the two-tier I-880 freeway in Oakland collapsed. The freeway collapsed exactly where it passed over soft, unconsolidated sediments, which amplified the shaking and overstressed the structure.

Figure 11.6. Air-photograph composite of San Francisco. (Image courtesy of the University of California, Santa Barbara Map and Image Library)

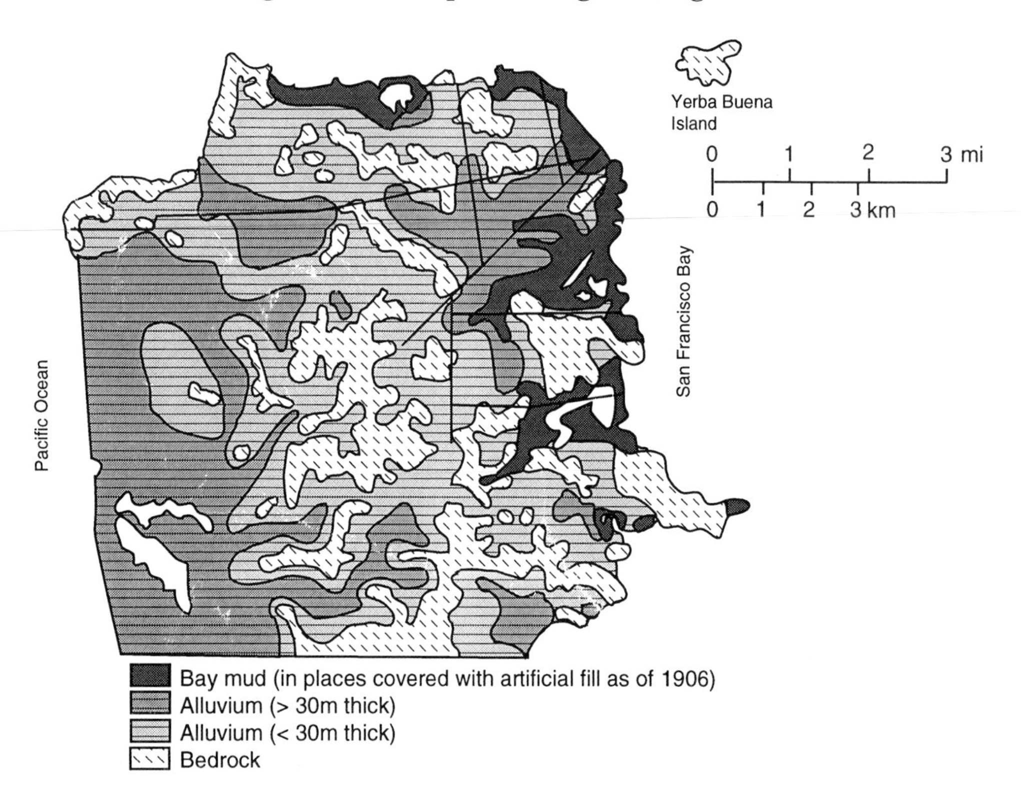

Figure 11.7. Surficial geology of San Francisco. (After Borcherdt, 1975)

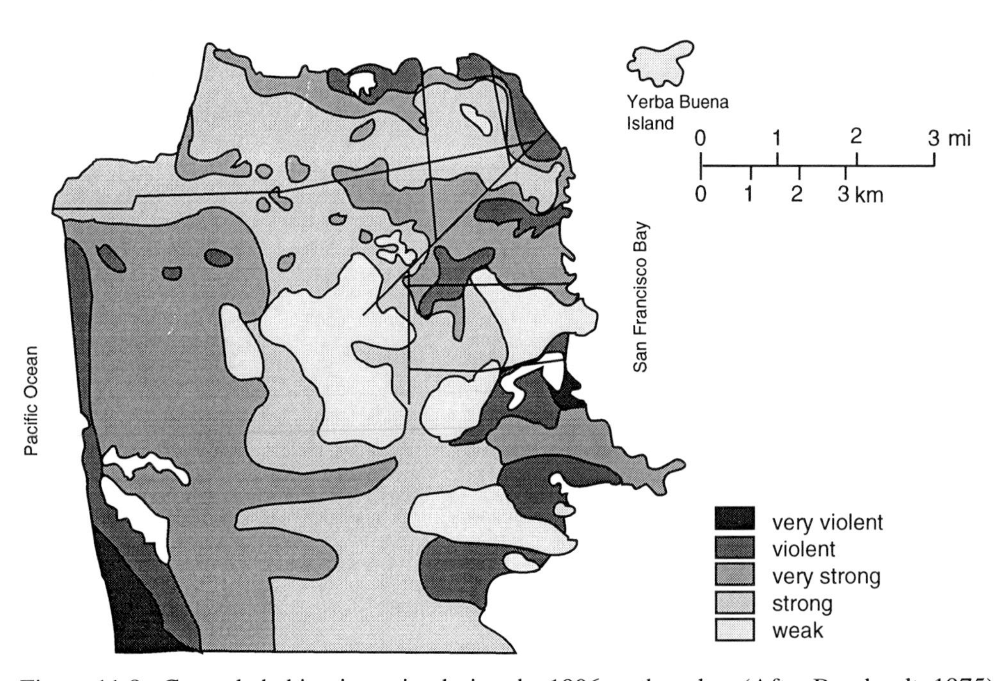

Figure 11.8. Ground-shaking intensity during the 1906 earthquake. (After Borcherdt, 1975)

The questions that follow ask you to evaluate the seismic hazard in San Francisco based on the information presented in this chapter.

7) Figure 11.7 shows a generalized map of the surficial geology in the San Francisco area and Figure 11.8 shows the distribution of ground shaking intensity during the 1906 earthquake. Compare these two figures. How does the intensity of the ground shaking correspond to the different kinds of earth materials found in the San Francisco area?

8) Using Figure 11.7, color the areas of potential material amplification in red, and mark areas of liquefaction by cross-hatching.

9) Indicate in blue pencil the areas in Figure 11.7 which you would recommend as a construction site for a building over 14 stories tall.

10) Using Equation 11.2, calculate the ground acceleration (in g) for a building at the corner of California and Market streets in San Francisco (Figure 11.6) for:

a) a 6.5 magnitude earthquake at 5 km distance.

b) a 7.5 magnitude earthquake at 5 km distance.

c) a 6.5 magnitude earthquake at 2 km distance.

11) Which parameter (earthquake magnitude or distance to the epicenter) has a greater effect on ground acceleration?

Seismic Shaking and Earthquake Engineering

In the following questions, you will calculate the base shear for a small wood-frame building which houses a retail store. The store occupies a three-story building, located at the corner of California Street and Market Street in San Francisco (see Figures 11.5 and 11.6). The building's total dead weight is 33,000 kg.

12) Using Equation 11.3 and Figures 11.1 and 11.7, calculate the actual base shear (in Newtons) imposed on the building during an intensity VII earthquake.

13) Using Equation 11.4, calculate the total base shear (in Newtons) that the building is expected to resist during an earthquake. For this kind of structure in this location:

$Z = 0.40$
$I = 1.0$
$C = 2.75$
$Rw = 8$

14) Although the results for both Questions 12 and 13 represent the estimates of base shear that would be imposed on this building, the results themselves are very different. Explain why the result obtained from Equation 11.4 is higher than the number obtained from Equation 11.3?

BIBLIOGRAPHY

Aitken, F. and E. Hilton, 1906. A History of the Earthquake and Fire in San Francisco. The Edward Hilton Co.: San Francisco.

Borcherdt, R.D. (ed.), 1975. Studies for seismic zonations of the San Francisco Bay region. U.S. Geological Survey Professional Paper 941-A.

Donovan, N.C., 1973. A statistical evaluation of strong motion data including the Feb. 9, 1971, San Fernando earthquake. Proc., 5WCEE, Rome, Italy, 1: 1252-1261.

Howell, B.F. Jr., 1973. Average regional seismic hazard index (ARSHI) in the United States. In D.E. Moran, J.E. Slosson, R.O. Stone, and C.A. Yelverton (eds.), Geology, Seismicity and Environmental Impact. Association of Engineering Geologists. University Publishing: Los Angeles.

Kehev, A.E., 1988. General Geology for Engineers. Prentice Hall, Englewood Cliffs, NJ.

Lagorio, H.J., 1990. Earthquakes: An Architect's Guide to Nonstructural Seismic Hazard. John Wiley & Sons: New York.

Leeds, D.J., 1973. The design earthquake. In D.E. Moran, J.E. Slosson, R.O. Stone, and C.A. Yelverton (eds.), Geology, Seismicity and Environmental Impact. Association of Engineering Geologists. University Publishing: Los Angeles.

Rahn, P.R., 1986. Engineering Geology: An Environmental Approach. Prentice Hall, Englewood-Cliffs, NJ.

Seed, H.B., 1972. Soil conditions and building damage in the 1967 Caracas earthquake, Journal of the Soil Mechanics and Foundations Division, American Society of Civil Engineers, SM-8: 787-806.

Seekins, L.C., and J. Boatwright, 1994. Ground motion amplification, geology and damage from the 1989 Loma Prieta earthquake in the city of San Francisco. Bulletin of the Seismological Society of America, 84: 16-30.

Wiegel, R.L., 1970. Earthquake Engineering. Prentice Hall, Englewood Cliffs, NJ.

Acknowledgements: This exercise was written by Susann Pinter, modified in part from Lagorio (1990).

EXERCISE 12

HYPSOMETRY AND CONTINENTAL TECTONICS

Supplies Needed

- calculator

PURPOSE

Hypsometry refers to the distribution of land area at different elevations. You can see that a high plateau that is cut by a few deep gorges is very different from a drainage basin with a well-developed dendritic stream network. Hypsometry provides the tools for measuring these differences and describing them quantitatively. The purpose of this exercise is to use hypsometry to evaluate how continental-scale tectonic processes shape the landscape.

INTRODUCTION

The Earth's land surface exists in a sensitive balance between tectonic forces and erosional forces. Erosion is controlled by climate and by the relief of the landscape, and tectonic processes have a major effect on relief. Where tectonic uplift and deformation are absent, erosional processes often predominate. After a sufficiently long period of time, the result is a subdued landscape with low elevation and little relief. Where uplift and deformation predominate over erosion, elevation and relief increase. This principle was the basis of W.M. Davis' *Cycle of Erosion,* in which he hypothesized that a surge of tectonic activity creates a landscape with "youthful" characteristics. Subsequent predominance of erosional processes reshapes the surface, forming "mature" landscapes, followed eventually by "old age" landscapes. Geomorphologists now recognize that Davis' model, with its mysterious surges of active tectonics, was excessively simplistic. Tectonic activity is no longer mysterious, but is explained by the interactions of the Earth's lithospheric plates. Landforms and landscapes in the different plate-tectonic settings reflect different tectonic processes at work. In this exercise, you will look at continental-scale landscapes, and you will use hypsometry to identify the regional effects of tectonic activity.

South America

The South American continent covers an area of about 18,700,000 km^2, including vast areas of Amazon rainforest and towering peaks of the Andes Mountains. Like North America, the western margin of the continent is an active plate boundary, while the east coast is a passive margin. The western margin of South America is the world's longest ocean-continent subduction zone. Subduction began in the early Jurassic period (around 200 million years ago). South America was then part of the supercontinent, Pangea. The modern Andes reflect shortening, compression, and volcanic activity that resulted from the subsequent 200 million years of subduction. From the Pacific Ocean eastward, the Andes rise from a narrow coastal plain to peaks that exceed 6,000 m in elevation. Along much of the range's length, the Andes consist of two distinct mountain chains: the Western Cordillera and the Eastern Cordillera, separated by the Altiplano ("high plain"). East of the Eastern Cordillera is the Subandean zone, foothills of the Andes that are up to 2000 m high. The Subandean zone is a Pliocene-age belt of folding and thrusting.

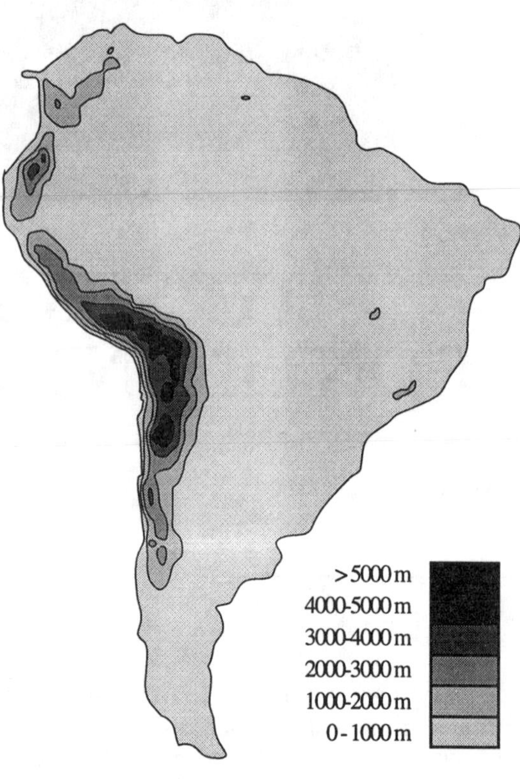

Figure 12.1. Generalized topography of South America.

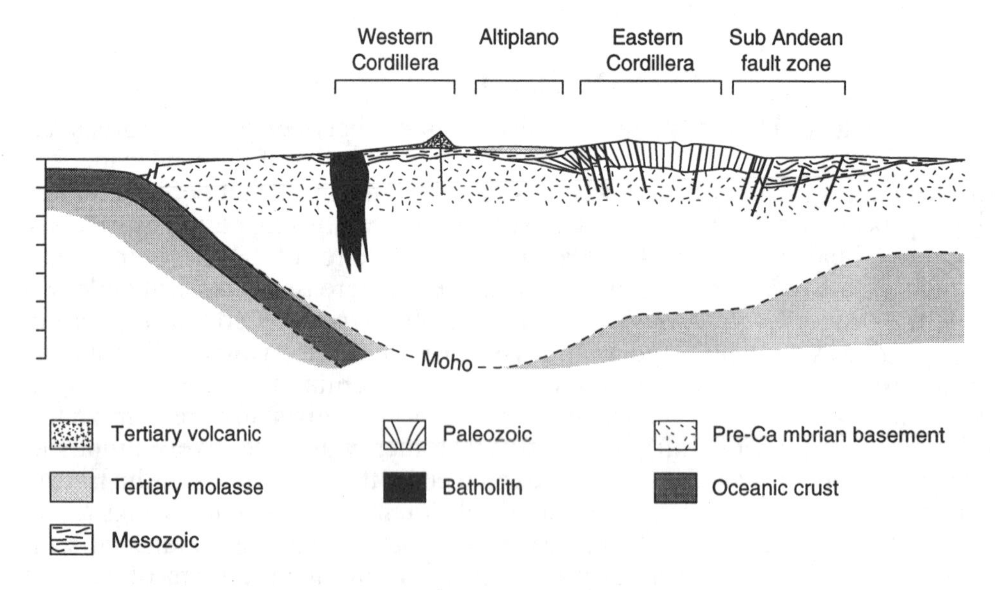

Figure 12.2. Cross section through the Peruvian Andes. (After Cobbing and Pitcher, 1972).

Figure 12.1 is a highly generalized illustration of the South American landscape. The cost of small map scale is loss of detail. The contour lines represent 1000 m steps in elevation. The hypsometric data in Table 12.1 is more detailed. It includes the total area between 200 m elevation steps. The last three columns in the table contain variables that are defined in Figure 12.4 and that you'll use in the next section.

Table 12.1. Hypsometry (elevation-area distribution) of South America.

A	B	C	D	E	F	G
from (m)	to (m)	Area (1000 km^2)	% of total area	a (1000 km^2)	$^h/_H$	$^a/_A$
0	200	2658.3	14.2	18679	0.00	1.00
200	400	8631.4	46.2	16021	0.03	0.86
400	600	3002.3	16.1	7390	0.06	0.40
600	800	1105.5	5.9	4387	0.09	0.23
800	1000	897.3	4.8	3282	0.11	0.18
1000	1200	323.1	1.7	2384	0.14	0.13
1200	1400	191.5	1.0	2061	0.17	0.11
1400	1600	206.7	1.1	1870	0.20	0.10
1600	1800	210.9	1.1	1663	0.23	0.09
1800	2000	117.1	0.6	1452	0.26	0.08
2000	2200	115.7	0.6	1335	0.29	0.07
2200	2400	88.3	0.5	1219	0.32	0.07
2400	2600	104.7	0.6	1131	0.34	0.06
2600	2800	131.4	0.7	1026	0.37	0.05
2800	3000	162.5	0.9	895	0.40	0.05
3000	3200	97.6	0.5	733	0.43	0.04
3200	3400	55.1	0.3	635	0.46	0.03
3400	3600	89.9	0.5	580	0.49	0.03
3600	3800	115.9	0.6	490	0.52	0.03
3800	4000	23.8	0.1	374	0.55	0.02
4000	4200	69.8	0.4	350	0.57	0.02
4200	4400	93.8	0.5	280	0.60	0.02
4400	4600	59.3	0.3	187	0.63	0.01
4600	4800	46.4	0.2	127	0.66	0.01
4800	5000	46.7	0.3	81	0.69	0.00
5000	5200	0.0	0.0	34	0.72	0.00
5200	5400	11.7	0.1	34	0.75	0.00
5400	5600	11.1	0.1	23	0.78	0.00
5600	5800	11.4	0.1	11	0.80	0.00
5800	6960	0.0	0.0	0	1.00	0.00

The simplest way to analyze elevation information such as in Table 12.1 is by using a **histogram**, as shown in Figure 12.3 below.

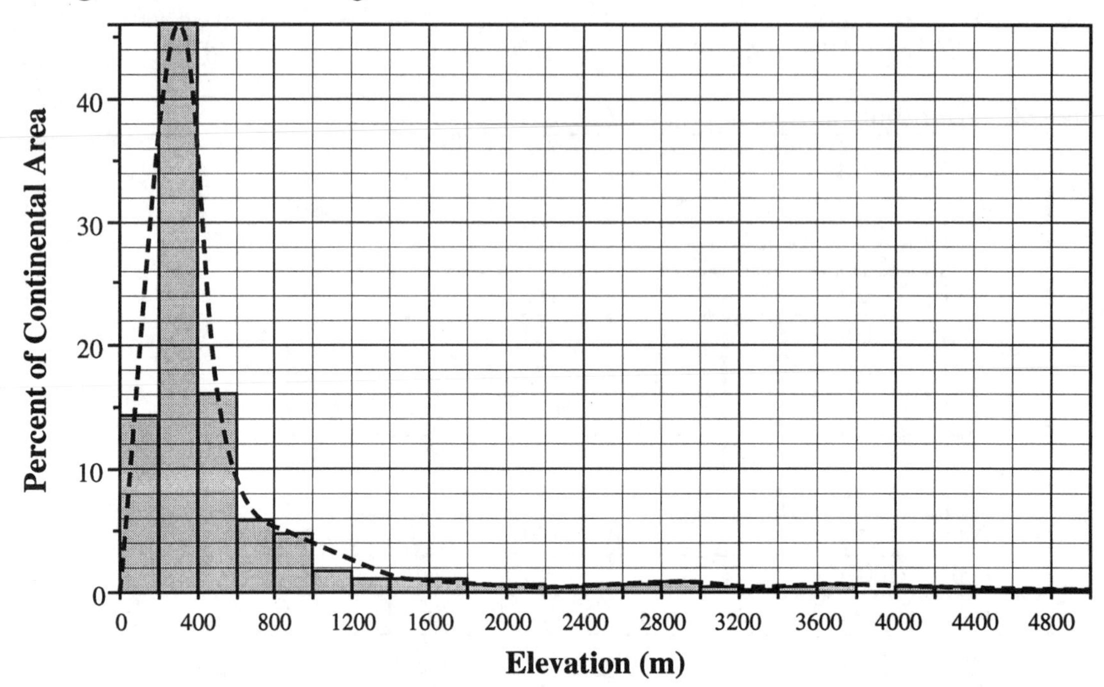

Figure 12.3. Histogram showing distribution of South American land area at different elevations. This figure shows one way to graphically illustrate the data in Table 12.1.

Figure 12.3 illustrates that a very large portion of South America (over 46%) occurs at a relatively low elevation, between 200 and 400 m. At the same time, the distribution also includes a much smaller proportion of the continent (about 4%) that rises from 3000 m in elevation to 5800 m and higher. Examine the two figures and the table on the previous page and answer the following questions about the topography of South America.

1) Looking at the map of South America (Figure 12.1), to what part of the continent does the sharp peak in the histogram between 200 and 400 m correspond?

2) What does the long "tail" on the right side of the histogram, from about 1200 m to 5800 m, represent?

3) The highest peak in South America is Mt. Aconcagua in Argentina, with an elevation of 6960 m. Why does Table 12.1 show no area higher than 5800 m? Another way to ask this question is: Would it be useful to extend Figure 12.3 all the way to 7000 m?

The Hypsometric Curve

Quantitative analysis of topography can be done with landscapes at any scale, from a single drainage basin to the entire planet. Comparing the results from different areas, however, can be difficult without some technique for removing the effects of scale. The *hypsometric curve* is a graph of area-altitude distribution that is *dimensionless,* meaning that it factors out both the total size of the area being studied and the total amount of relief. The hypsometric curve is a plot of *relative height* (h/H) versus *relative area* (a/A), as shown in the diagram at the top of the next page. By dividing the height of a given point (h) by the total relief (H), the relative height of any point is simply a number between 0.0 and 1.0. At any given point, the parameter **a** is a measurement (in km^2) of the area *that is higher in elevation* than that point. Standing on the highest mountain peak on a continent, nothing is higher than you (a=0.0); standing at sea level, all of that continent is higher. As was done for height (a=1.0), **a** is divided by the total area of the landscape (A) to give a dimensionless value between 0.0 and 1.0. Using the hypsometric curve, you can compare different areas in order to study the effects of different bedrock types, different plate-tectonic settings, or the balance between tectonics and erosion.

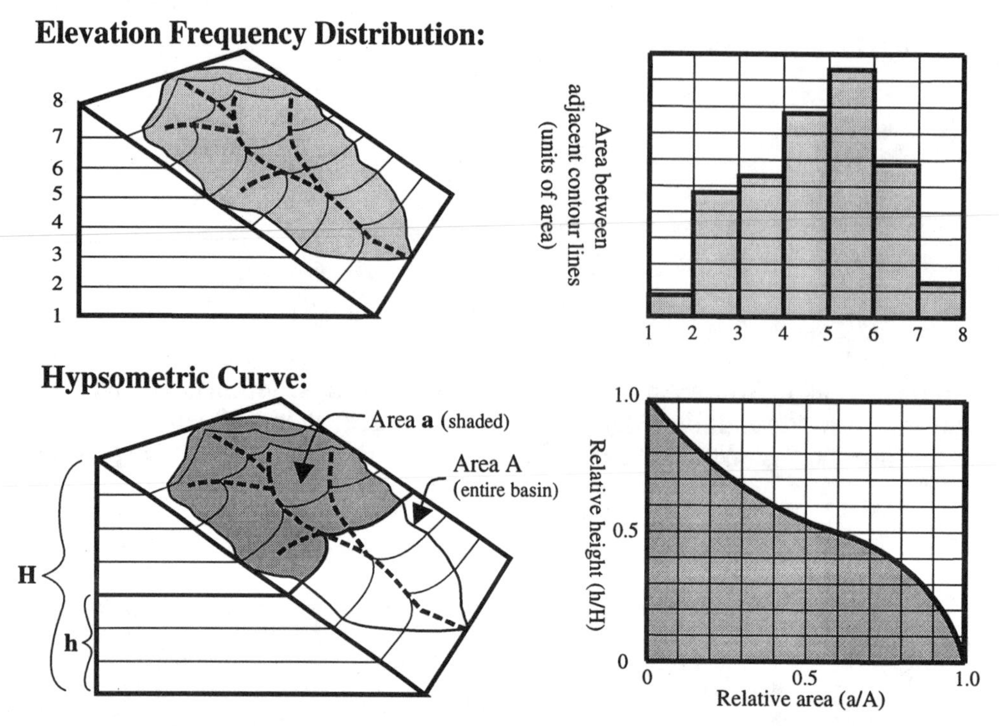

Figure 12.4. The relationship between the elevation-frequency distribution for a landscape (in this case, one drainage basin) and the hypsometric curve. The hypsometric curve is constructed by measuring the area (a) that is higher than any given elevation (h) on the landscape.

The hypsometric curve of South America is in Figure 12.5 to the right. It shows that the highest peaks (above 1200 m or more) comprise a very small portion of the continent's total area (the curve above 1200 m hugs the left side of the graph). As shown on the shaded relief map of South America earlier in this exercise, the great majority of the continent lies at elevations lower than 1000 m. Answer the questions on the following page.

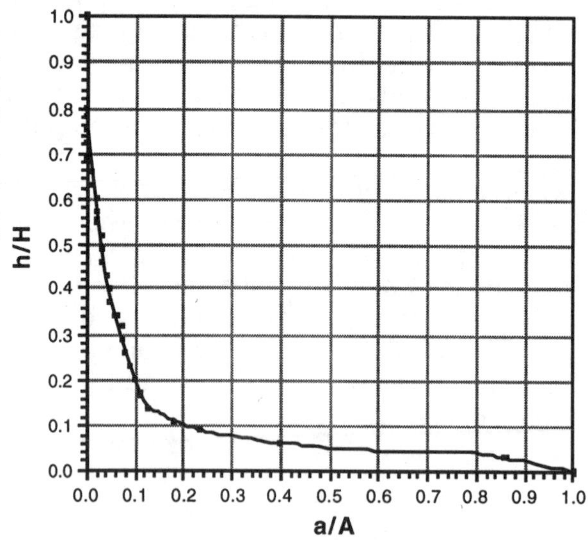

Figure 12.5. Hypsometric curve for South America.

4) What is the median elevation (in meters) of South America (50% of the continent higher and 50% lower)? (Remember that the highest point on the continent is 6960 m)

5) Use Figure 12.5 to determine what percent of South America is higher than 1740 m.

Table 12.2. Hypsometry of North America.

A	B	C	D	E	F	G
from (m)	to (m)	Area (1000 km^2)	% of total area	a (1000 km^2)	$^h/_H$	$^a/_A$
0	200	6263	23.71	26,417	0.00	1.00
200	400	7276		20,154		
400	600	3924	14.86	12,877	0.06	0.49
600	800	2201		8953		
800	1000	1906	7.22	6753	0.13	0.26
1000	1200	1204		4846		
1200	1400	575	2.18	3642	0.19	0.14
1400	1600	508		3067		
1600	1800	934	3.54	2559	0.26	0.10
1800	2000	757		1625		
2000	2200	454	1.72	868	0.32	0.03
2200	2400	196		414		
2400	2600	72	0.27	218	0.39	0.01
2600	2800	75		146		
2800	3000	50	0.19	71	0.45	0.00
3000	3200	22		22		
3200	6194	0	0.00	0	1.0	0.00

Figure 12.6. Generalized topography of North America.

Hypsometry and Continental Tectonics

North America

The remainder of this exercise will allow you to work with the hypsometric data for North America, plotting the elevation-frequency distribution and the hypsometric curve. The elevation information for North America is shown in Table 12.2.

6) Your first task will be to calculate the percent of the continent that falls within each elevation interval (Column D in Table 12.2). You'll need to note that the total area of North America is 26,417,000 km^2 (26,417 * 1000 km^2). Every second value is calculated for you. This kind of laborious calculation is best done on computer, using a spreadsheet program for example, but for the relatively small number of intervals used here, a calculator is fine.

7) Using the values you just calculated, plot the elevation frequency histogram for North America. Go back to the histogram for South America (Figure 12.3), and plot the North American data on top of it.

8) If you plotted your data correctly, you should see that North America lacks the low-elevation spike and the high elevation tail of the South American distribution. Why is that?

9) In preparation for plotting the hypsometric curve, you will need to calculate the values for relative height (h/H; Column F) and relative area (a/A; Column G). When calculating **h/H**, you will need to know that the highest point in North America is Mt. McKinley in Alaska, with an elevation of 6194 m. To be consistent, use the *minimum* elevation (Column A) for **h**. As before, every second value already has been calculated for you.

10) Plot the hypsometric curve for North America. Go back to Figure 12.5 (the hypsometric curve for South America), and add the North American data. Strictly using W.M. Davis' model of landscape evolution, which continent looks more "youthful"?

The graph in Figure 12.7 to the right is a modified hypsometric curve. Like the conventional hypsometric curve, area is plotted as relative height. Unlike the conventional hypsometric curve, elevation is left in absolute units (meters). Curves for Asia, South America, and North America are shown. You can see not only that Asia has a higher maximum elevation than North and South America, but also that a relatively large portion of the Asian continent lies at high elevations. Answer questions 6, 7, and 8 on the next page:

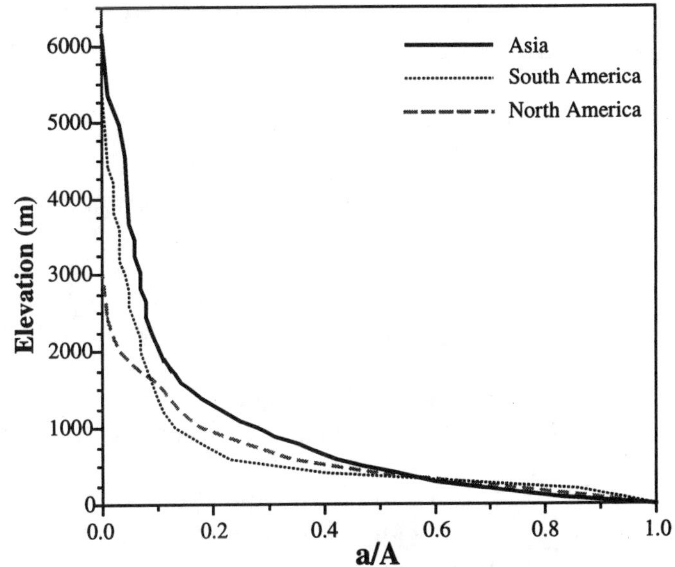

Figure 12.7. Modified hypsometric curves for Asia, South America, and North America.

11) Most of North America, even areas with high elevations, are not currently undergoing rapid uplift or mountain-building. Which graph illustrates this fact better: the conventional hypsometric curve or the modified curve?

12) Look back at the elevation histogram for North America. You should see a bulge in the distribution between around 1600 and 2000 m. To what portion of the North American landscape does this elevation range correspond?

13) The North American hypsometric curve crosses the South America curve at around 1800 m in elevation and stays above it until around 600 m, meaning that a larger portion of North America occupies intermediate elevations. How does this fact reflect the tectonic history of North America?

BIBLIOGRAPHY

Cobbing, E.J., and W.S. Pitcher, 1972. The segmented coastal batholith of Peru: Its relationship to volcanicity and metallogenesis. Earth Science Reviews, 18: 241-251.

Cogley, J.G., 1985. Hypsometry of the continents. Zeitschrift für Geomorphologie Supplementbände, 53.

Harrison, C.G.A., K.J. Miskell, G.W. Brass, E.S. Saltzman, and J.L. Sloan II, 1983. Continental hypsometry. Tectonics, 2: 357-377.

James, D.E., 1973. The evolution of the Andes. In F. Press and R. Siever (eds.), Planet Earth. W.H. Freeman and Company: San Francisco.

Mégard, F., 1987. Structure and evolution of the Peruvian Andes. In J-P. Scaer and J. Rodgers (eds.), The Anatomy of Mountain Ranges. Princeton University Press: Princeton, NJ.

Ramos, V.A., 1989. The birth of southern South America. American Scientist, 77: 444-450.

Strahler, A.N., 1952. Hypometric (area-altitude) analysis or erosional topography. Geological Society of America Bulletin, 63: 1117-1142.

Thornbury, W.D., 1965. Regional Geomorphology of the United States. John Wiley and Sons, New York, 609 pages.

GLOSSARY

accelograph: instrument that measures ground acceleration during seismic shaking

active fault: a fault that has moved within a given period of time, typically the past 10,000 years

aftershock: an earthquake that follows and is less powerful than the main shock

aggradation: in a *fluvial system,* the accumulation of sediments, in response to a rise in *base level* or other causes

alluvial fan: a conical-shaped landform that characterizes mountain fronts of arid and semi-arid regions

alluvium: loose sedimentary material deposited by rivers or streams

aseismic: describes an event or process that occurs without accompanying earthquake activity

asthenosphere: plastic layer of the upper mantle that lies beneath the *lithosphere*

Asymmetry Factor (AF): a geomorphic index used to detect active tilting

axial surface: on a fold, the surface that bisects the angle between the two fold limbs

balanced cross section: a geologic cross section in which strata are parallel, and individual layers maintain uniform or uniformly-varying thickness

base level: the lowest elevation that a specific fluvial system drains to; the concept includes both local base level and ultimate base level (usually sea level)

body waves: seismic waves that travel through the interior of the Earth

brittle behavior: when a material responds to an applied stress by fracturing

buried reverse fault: a compressional fault that did not rupture the ground surface when it was active

characteristic earthquake: an earthquake that strikes a given fault zone with approximately the same magnitude and other characteristics at approximately equal intervals

coastal terrace: (sometimes called an *uplifted coastal terrace*) a set of coastal landforms, typically either a *wave-cut platform* or a coral reef complex, that has been uplifted above the modern shoreline

colluvial wedges: a deposit of colluvium at the base of a slope, thickest near the slope and progressively thinner farther away

colluvium: loose sedimentary material deposited by gravity-driven processes, usually at the base of a slope

conditional probability: probability that a given event will occur in a specified period of time

contour interval: the vertical distance separating two adjacent *contour lines*

contour line: a line of equal surface elevation on a topographic map

coseismic: describes an event or process that coincides with an earthquake

creep: 1) slow downslope movement of *colluvium,* 2) slip on a fault without measureable earthquakes

cross-cutting relationships: the principle that: if geological feature A cuts feature B, then B must have formed before A

deflected drainage: a stream that follows a strike-slip fault along some or all of its length

degradation: in a *fluvial system,* the removal of sediments or erosion of the channel, in response to a fall in *base level* or other causes

dendritic drainage pattern: "finger-like" pattern of streams associated with homogeneous bedrock and gentle slopes

denudation: regional erosion of the surface

diffusion equation: a mathematical expression that is used to quantitatively model *fault-scarp degradation*

dip: the maximum slope angle on a sloping surface

drainage basin: the area in which all rain that falls exits through the same stream

drainage divide: the line (usually a ridge or crest) that separates two *drainage basins*

earthquake: a sudden motion or trembling in the Earth caused by the abrupt release of strain on a *fault*

elastic behavior: deformation that is recovered fully and instantaneously when the driving force is removed

elastic rebound model: model in which faults accumulate elastic *strain,* and earthquakes represent the sudden release of that strain

emergence: motion of the land up relative to sea level, such that the coastline advances oceanward

epicenter: the point on the surface of the Earth directly above the *focus* of an earthquake

erosion: general term describing the processes of *weathering* and transport of sediment

fault scarp: a steep slope formed by a fault rupturing the surface

fault-scarp degradation: the systematic process by which a *fault scarp* goes from a sharp, steep slope shortly after formation to a more curved, gently-sloping shape after a number of years

fault zone: a related group of faults in a subparallel belt or zone

fault-bend fold: a fold formed by a change in dip on an underlying fault

fault-propagation fold: a fold that forms around the tip of a fault that does not rupture the ground surface (see *buried reverse fault*)

fault: a break in the Earth's crust on which rupture occurs or has occurred in the past

floodplain: flat land or valley floor that borders a stream or river, formed by migration of meanders and/or periodic flooding

fluvial geomorphology: the study of river processes and landforms caused by river processes

fluvial system: a river or stream

focus: the location within the Earth at which an earthquake originates

fold-and-thrust belt: a zone characterized by faults and folds that reflect active compression

footwall: the side of a fault that lies beneath the inclined fault plane

foreshock: an earthquake that precedes and is less powerful than the main shock

geologic structures: features produced in rock by movement after deposition or formation

ground acceleration: a quantitative measurement of the intensity of seismic shaking (usually given as a percent of the acceleration of gravity)

half-life: time interval during which exactly one-half of the unstable parent isotopes originally present in a system decay into their daughter products

hanging wall: the side of a fault that lies above the inclined fault plane

Holocene: the last 10,000 years

hypsometric curve: a graphical representation of the elevation distribution of a given landscape

hypsometric integral: area under the *hypsometric curve*

hypsometry: measurement and analysis of the relationship between elevation and area

incision: local or regional erosion by streams, typically causing an increase in *relief*

intensity (of an earthquake): a relative measurement of the strength of shaking at any given location. Intensity generally decreases with increasing distance from the epicenter.

interseismic: describes an event or process that occurs between major earthquakes

intraplate earthquakes: earthquakes that occur in the interior of a lithospheric plate, away from any plate boundary

isostasy: the principle by which thicker, more buoyant crust stands topographically higher than thinner, denser crust

isostatic compensation: the principle that high-standing topography is underlain and supported by a thick and/or low-density crustal root

isotopes: elements that have the same number of protons but differing numbers of neutrons. Unstable isotopes are subject to radioactive decay

kink: sharp bend in a rock layer or unit

landform: a discrete element of the landscape, such as a hill, a terrace, or an alluvial fan

liquefaction: transformation of water-saturated sediments from a solid to a liquid state in response to shaking

lithosphere: the upper portion of the Earth, consisting of the crust and upper portion of the mantle, that is characterized by brittle behavior

magnitude (of an earthquake): an absolute measurement of the energy of a given earthquake

material amplification: a local increase in the intensity of seismic shaking caused by near-surface material (usually loose sediments)

maximum credible earthquake: the largest earthquake likely to be generated by faults in a given area

Mercalli Scale: a system for estimating earthquake *intensity*

moment magnitude: a system for measuring earthquake *magnitude* based on the total energy released by the earthquake (also see *seismic moment*)

morphological dating: estimating the age of a landform based on its shape, usually estimating the amount of erosion that has occurred since the landform was formed

morphometric index: a quantitative parameter measured of calculated from surface topography
morphometry: quantitative measurement and analysis of topography
mountain front: steep escarpment that marks the boundary between mountainous topography and relatively flat topography
normal fault: a fault across which there is extension
numerical age control: estimates of the age of a material or feature in an absolute number of years (as opposed to *relative age control*)
offset stream: a stream the channel of which is displaced across a strike-slip fault
P-wave: compressional ("push-pull") seismic waves
paleoseismology: the study of earthquakes that occurred in the geologic past
piercing points: two points on opposite sides of a fault that were originally connected, but were offset by one or more ruptures along the fault
plastic behavior: a permanent change in the shape of a material after a force is applied to it
plate tectonics: the theory that the Earth's lithosphere is subdivided into a series of discrete plates that each move relative to the others
Pleistocene: the period of geologic time from about 1.65 million to 10,000 years ago. Much of the Pleistocene was characterized by the growth and decline of glaciers in many areas.
postseismic: describes an event or process that occurs shortly after an earthquake
preseismic: describes an event or process that occurs shortly before an earthquake
pressure ridge: a hill along a strike-slip fault zone formed by upwarping at *restraining bends* or between two different strands of the fault
progradation: oceanward advance of a coastline, usually by deposition of sediment
Quaternary: the latest period of geologic time up to and including the present. The Quaternary includes the *Pleistocene* and the *Holocene* , and ranges from approximately 2 million years ago to the present.
radiocarbon dating: a method that estimates the absolute age of a sample based on the ratio of radiogenic carbon (^{14}C) to stable carbon
recurrence interval: the average period of time between major earthquakes on a given fault (see *characteristic earthquake*)
relative age control: estimates of the age of a material or feature compared with other features (as opposed to *numerical age control*)
relative spacing: a method for determining the ages of a sequence of *coastal terraces* based on their present-day elevations and knowledge of Quaternary sea-level history
releasing bend (or step): a bend (or step) in a strike-slip fault that causes localized extension
elief: generally, the "ruggedness" of the topography; specifically, the highest elevation minus the lowest elevation in a given area
response spectra: in earthquake engineering, it is the relationship between seismic-wave period and ground shaking
restraining bend (or step): a bend (or step) in a strike-slip fault that causes localized compression
etrodeformation: interpretation of a geologic cross section with the goal of understanding its geometry before deformation occurred
retrogradation: landward retreat of a coastline, usually by erosion
reverse fault: a fault across which there has been convergence
Richter magnitude: (also called local magnitude) a system for measuring earthquake *magnitude* based on the maximum amplitude and period of seismic waves recorded by a seismograph at a set distance from the earthquake *epicenter*
rupture: breakage of a material under stress
S-wave: shear ("side-to-side" motion) seismic waves
sag pond: a pond along a strike-slip fault zone formed by downwarping between two different strands of the fault
sand boil (also called a "sand blow," or a "sand volcano"): sand extruded during seismic shaking, caused by high fluid pressure and *liquefaction*
scarp: (short for "escarpment") a slope steeper than the surrounding topography; related to a change in material, process, or geomorphic history
sea cliff: on a *coastal terrace* or a modern coastline, it is the steep slope cut by wave action at its base
segmentation: subdivision of a fault zone into smaller units with discrete rupture histories
seismic gap: a portion of a fault zone, between two areas that have ruptured in historical or recent time, that has not ruptured
seismic moment: a measurement of the total amount of energy released during an earthquake, measured in dyne*cm or equiavalent units
seismic risk: an estimate of the likelihood and the potential damage of an earthquake in a given area
seismic waves: energy released from a fault rupture, subdivided into body waves (P-waves and S-waves) and surface waves (Rayleigh waves and Love waves)

seismic zoning: legal definition of land as appropriate or inappropriate for different uses based on proximity to active faults, presence of material that may amplify shaking, etc.

seismogram: graphical or digital record of *seismic waves,* recorded on a *seismograph*

seismograph: instrument for measuring *seismic waves;* the record of seismic waves itself is called a *seismogram*

seismology: the study of *seismic waves,* specifically with the purpose of better understanding earthquakes and the interior of the Earth

seismometer: instrument used to measure seismic waves

shoreline angle: on an erosional coastline, the line at which the wave-cut platform meets the seacliff

shutter ridge: a ridge offset by a strike-slip fault such that the ridge below the fault is juxtaposed against the gully above the fault

slip rate: long-term rate of motion on a fault or fault zone

slope angle: the steepness of a slope expressed in degrees or radians

slope gradient: the steepness of a slope expressed as the unitless ratio, rise ÷ run

stereocope: optical instrument for simultaneously viewing two photographs and thus producing a three-dimensional perspective

strain: deformation resulting from *stress*

strain partitioning: observation that in areas with oblique strain, horizontal and vertical deformation often occur on distinct and separate structures

Stream Length-Gradient Index (SL): a geomorphic index used to identify possible areas of tectonic activity

stress: a force applied to an object or material

strike: on an inclined surface (often a sedimentary later) the orientation of a line of equal elevation

strike-slip fault: a fault along which two tectonic blocks slide laterally past each other

submergence: motion of the land down relative to sea level, such that the coastline retreats landward through time

subsidence: downwarping of an area of the Earth's surface

superposition: the principle that: if layer A overlies layer B, then B formed before A

surface waves: seismic waves that travel along the solid surface of the Earth

tectonic geomorphology: 1) the study of landforms shaped by tectonic process; 2) application of geomorphic principles to reveal the presence, pattern, or rates of tectonic processes

tectonics: processes that deform (move) the Earth's crust, and the structures and landforms that result from those processes

terrace: a topographic bench that is not longer actively forming, typically near an active stream or coastline. The term is applied to both the flat surface of the terrace (the "tread") and the slope below (the "riser")

threshold: a critical transition point in a system, such as the maximum amount of change that a system can absorb before its dynamic equilibrium becomes unbalanced

thrust fault: a type of *reverse fault* which is less steep than 45°

tilting: process by which a horizontal surface acquires a slope (usually without *warping)*

topographic profile: vertical cross section that shows changes in the elevation of the surface along a single line

transport-limited slope: a hillslope on which erosion is limited only by the rate of sediment transport (also see *weathering-limited slope*)

Transverse Topographic Symmetry Factor (T): a *geomorphic index* used to detect active *tilting*

travel-time curve: graph that shows the relationship arrival times of different *seismic-waves* as a function of distance from the *epicenter*

trunk stream: in a *drainage basin,* the principal or central stream

uplift path: on a graph of sea-level history (elevation versus age), it is the line that traces how an individual *coastal terrace* was formed and uplifted to its present elevation

volcanic tumescence: uplift caused by rising magma beneath the surface

warping: process by which a planar surface becomes folded

wave-cut platform: on a *coastal terrace* or a modern coastline, it is the subhorizontal surface cut by waves as well as by secondary biological and chemical processes

weathering: physical and chemical processes that break down rock at and near the Earth's surface

weathering-limited slope: a hillslope on which erosion is limited only by the rate of bedrock (or substrate) *weathering* (as opposed to a *transport-limited slope)*